Contents

Preface

Since 2003, when the fourth edition of this book was published, the world of estate agency has had to endure the cataclysmic effects of the "credit crunch" and a global recession. Activity in property markets, both residential and commercial, has slowed to a trickle compared to the heady days of the previous decade, and many agency firms have been forced to retrench, some to close altogether. As this edition goes to press, some commentators are claiming to see the first signs of an economic upturn, though opinion remains sharply divided as to whether these really are the green shoots of recovery or merely a summer-time twitch before another desperate winter.

Not surprisingly, given the huge fall in the number of property transactions taking place, there has been a reduction in estate agency-focused litigation since the previous edition. That is not to say that the courts have stood entirely idle: important decisions have been forthcoming on fees (notably commission made payable on exchange of contracts and the "effective cause" principle) and on the responsibility of a client whose conduct vis-à-vis the purchaser brings about the collapse of the sale. These rulings, together with others interpreting relevant legislation such as the Property Misdescriptions Act and the Unfair Terms in Consumer Contracts Regulations are all discussed in the text.

That said, the most significant (for estate agents) law-making activity in the past few years has taken place, not in the courtroom, but at Westminster. Home Information Packs (albeit in a somewhat emasculated form without the much-trumpeted Home Condition Reports) are now an established feature of the residential property market, and it is the agent who bears responsibility under the Housing Act 2004 to ensure that these are in place before a property comes to market. The government, while continuing to balk at the licensing of estate agents, has made membership of an "Ombudsman" scheme compulsory, at least in the residential sector. And, as in other areas of activity regarded by government as matters of consumer protection, the authorities with responsibility for the administration of the Estate Agents Act have seen a considerable strengthening of their enforcement powers.

The task of preparing this new edition has, as usual, been considerably lightened by sisters Audrey Andersson and Amy Boyle, whose combined efforts serve to protect the author from all that is practical or technical (and therefore uncongenial) in the publication process. And to Sandi, as ever, my love and gratitude for help in so many ways.

John Murdoch
St George's Day 2009

Table of Cases

G

J

K

L

W

Table of Statutes

Table of Statutory Instruments

Introduction to Agency Law

The law of agency is concerned with the activities of three characters:[1] an agent who acts on behalf of a *principal* or *client* in dealing with a *third party*. If these three characters are visualised as standing at the points of a triangle, then the sides of that triangle represent the legal relationships between the parties. This in turn serves to emphasise an important truth, namely, that the law of agency deals, not just with one legal relationship (that of principal and agent), but with all three.

There is indeed something to be said for the view that the most important aspect of agency law is the relationship between the principal and the third party. After all, the primary purpose of appointing an agent is to permit the principal to enter into transactions with third parties, without necessarily having to act in person. The fundamental question which has to be addressed, therefore, is how far the agent's acts are to be regarded as equivalent to acts carried out personally by the principal. Other questions (such as the agent's legal position *vis-à-vis* the principal and the third party) are also important, but they may perhaps be regarded as ancillary to the central issue.

Some idea of the kind of legal problems with which the law of agency deals may be obtained by considering the following hypothetical situations, each of which concerns an agent representing a principal in dealing with a third party:

---

1   Or four, in cases where a *sub-agent* is involved.

- The agent fulfils the principal's instructions to the letter, making the existence of the agency clear to the third party. Here a legal relationship comes into existence between the principal and the third party; the agent drops out of the picture, reappearing only to claim from the principal whatever is due under the agency agreement.

- The agent again makes a contract with the third party, but this time exceeds whatever authority has been given by the principal. Here the principal, who is not bound by the agent's acts, can effectively disclaim all liability on this contract. If the third party is to obtain any form of redress, this can only come from the agent.

- There are some circumstances, which will be explained later, where the principal can be legally bound by acts of the agent which have not been expressly authorised; indeed, they may have been specifically prohibited. In such a case, the principal, whose legal position *vis-à-vis* the third party has been wrongfully altered, may take action against the agent for breach of duty.

Agency and authority

If any one concept may be singled out as central to the law of agency, it is surely that of 'authority'. The scope of the authority given by the principal to an agent is crucial in determining whether or not the third party can hold the principal responsible for what the agent has done. It is also of great importance in deciding whether or not the agent is in breach of duty towards the principal. Again, where the potential personal liability of the agent towards a third party is in issue, the scope of the agent's authority is an important factor to be considered.

In recognising the importance of authority, it is vital to appreciate that the word takes on different shades of meaning in different contexts. As far as the third party is concerned, any act of the agent which is binding upon the principal may be said to have been 'authorised' in one way or another. In considering the agent's position, however, 'authority' bears a much more restricted meaning; an agent's primary duty is that of obedience to the principal, whose express instructions (as properly interpreted and with any appropriate implications) are paramount.

The first part of this chapter is concerned to identify the various kinds of authority which an agent may possess. As will be seen, these are not all conferred by the principal in any direct sense; they may arise by implication from the parties' actions, or even by operation of law. It is thus important that the types of authority, and the differences between them, are fully understood.

Express authority

Agency is a consensual relationship; indeed, it arises most commonly out of a contract entered into between an agent and a principal, under which the agent agrees to perform certain tasks in return for some reward. Even where there is no contract in the full legal sense, as where the agent undertakes to do something for nothing, most (though not all) of the normal incidents of the agency relationship will be present.

Where agency *is* based on a contract, it is subject to all the general rules of law which govern the validity of a contract, such as fraud, misrepresentation, mistake or illegality. In particular, persons such as minors or the mentally incompetent, whose capacity to make contracts is legally restricted, may only appoint agents to carry out such transactions as they could enter into for themselves,[1] and cannot in any event execute a valid power of attorney (that is, a deed conferring powers on an agent).[2] Furthermore, an agent's authority will be automatically terminated if an initially sane principal becomes mentally incompetent, although the third party may continue to rely upon the appearance of authority until notified of the principal's condition.[3]

Surprisingly, perhaps, an agent who lacks contractual capacity may still bring about a binding contract between the principal and a third party. In such circumstances, however, the agent will not be bound by the contract of agency and will not incur personal responsibility to the third party, in circumstances where an agent of full capacity would do so.[4]

As to the form of a contract of agency, the general principle is that, apart from certain statutes which require an agent's appointment to be

---

1   *Doyle* v *White City Stadium Ltd* [1935] 1 KB 110.
2   *Zouch d Abbott and Hallet* v *Parsons* (1765) 3 Burr 1794; *Daily Telegraph Newspaper Co* v *McLaughlin* [1904] AC 776.
3   See p 37.
4   *Smally* v *Smally* (1700) 1 Eq Cas Abr 6.

by deed[1] or in writing,[2] no specific formalities are necessary. Thus an agent instructed to execute a contract, which must itself be in writing, may be validly appointed by word of mouth.[3] However, an agent who is to execute a deed on behalf of the principal must receive authority to do so in the form of a power of attorney,[4] except where the agent executes the deed in the principal's name and presence and at the principal's express request.[5]

The scope of an agent's express authority naturally depends upon the terms of appointment. Where these are contained in a deed, they will be strictly construed.[6] In consequence, general words which appear to confer wide powers upon an agent may well be cut down by special words in another part of the instrument. In *Danby* v *Coutts & Co Ltd*,[7] for example, the plaintiff appointed by deed two persons to be his attorneys, with authority to borrow money on mortgage. The deed did not limit the duration of these appointments; however, it began with a recital that the plaintiff was going abroad for a time and required an attorney to act for him during his absence. When the two agents fraudulently executed a mortgage of the plaintiff's property, in order to raise money for their own purposes, it was held that this was not binding upon the plaintiff, since it was executed after the latter had returned to England. The court ruled that the recital governed the operative part of the deed, so as to limit the agents' powers to the time when the principal was abroad.

Legal interpretation of documents other than deeds is much less strict, far more regard being paid to the surrounding circumstances. As a result, an agent is allowed to exercise a certain amount of discretion for the benefit of the principal, although not to override any express instructions. This principle was applied somewhat generously in *Ireland* v *Livingston*,[8] where an agent had been instructed to purchase 500 tons of sugar, give or take 50 tons more or less. When the agent found that only 393 tons were available and duly purchased this

---

1   Eg Trustee Act 1925, s 25 (delegation of powers by trustee going abroad).
2   Eg Law of Property Act 1925, ss 53, 54 (creation of interests in land and declaration of trusts).
3   *Re Whitley Partners Ltd* (1886) 32 ChD 337.
4   *Steiglitz* v *Eggington* (1815) Holt NP 141; *Berkely* v *Hardy* (1826) 5 B & C 355.
5   *Ball* v *Dunsterville* (1791) 4 TR 313.
6   *Bryant, Powis & Bryant Ltd* v *La Banque du Peuple* [1893] AC 170, 177.
7   (1885) 29 ChD 500. See also *Jacobs* v *Morris* [1902] 1 Ch 816.
8   (1872) LR 5 HL 395.

quantity, the House of Lords held that the transaction lay within the agent's authority. On the other hand, where an agent was instructed to act 'in and about the purchase' of a colliery, it was held that this did not include authority to sign a binding contract of purchase.[1]

Implied authority

In a number of different situations, the law will imply authority for an agent to act on behalf of a principal. Before examining these situations, it is important to emphasise that the authority which they confer is as real as if it were expressly given. Unlike the doctrine of 'apparent' or 'ostensible' authority,[2] which permits a third party to rely on an appearance of authority created by the principal, implied authority also protects an agent who acts within its scope.

Of course, authority will not be implied in the face of express instructions to the contrary from the principal, and an agent who contravenes those instructions will be personally liable to the principal. However, the third party in such circumstances may continue to rely on the agent's apparent authority, unless and until notice of its exclusion or restriction has been given.

Implication from the parties' conduct

Where the very appointment of an agent is based upon the parties' actions, rather than upon their words, any resulting authority must by definition be implied. An example of this situation is *Sims* v *Landray*,[3] in which the defendant bid successfully for property at an auction. The auctioneer's clerk inserted the defendant's name into the memorandum of sale, while the defendant stood beside him without demur. It was held that the defendant must be taken to have authorised the clerk to sign the memorandum on his behalf.

Apart from these (admittedly unusual) cases in which the agency itself rests upon implication, an agent's express authority 'implies and includes a right to do all subordinate acts incident to and necessary for the execution of that authority'.[4] 'Implication' in this sense might

---

1   *Vale of Neath Colliery* v *Furness* (1876) 45 LJ Ch 276.
2   See p 9.
3   [1894] 2 Ch 318.
4   *Collen* v *Gardner* (1856) 21 Beav 540, 543. See also *Pole* v *Leask* (1860) 28 Beav 562, 574.

equally be regarded as an interpretation of the agent's express authority; either way, it is of a very limited nature, so that for example an agent authorised to sell land has no implied authority to receive the purchase price, although the agent can accept a deposit.[1]

Implication from the business context

The law has long taken the view that, where an agent of a particular type is appointed, both the agent and any third party with whom he or she deals should be able to assume (until notified to the contrary) that the agent has been given whatever authority is usual in the circumstances. This view has created a kind of implied authority known as 'usual authority', which arises in two rather different situations.

In the first place, an agent may be appointed to a position in the principal's business. Where this is so, it is assumed that the agent has whatever authority is usually conferred upon the holder of such a position. Thus, for example, an agent employed in a general capacity to manage the principal's estate may validly accept a notice to quit from a tenant.[2] By contrast, a notice to quit given by a tenant to a rent collector would not be valid, for it is not within the usual authority of such an agent to receive it.[3] A similar distinction may be found in relation to companies; a managing director has very wide implied powers to commit the company to contracts,[4] whereas the authority of the other individual directors is much more limited in scope.[5]

The second type of 'usual authority' consists of what is normal in the agent's own trade, business or profession. In this connection it is well established that a principal is bound by any trade custom which is held to be reasonable,[6] and even by an unreasonable custom which the principal is aware of and has agreed to.[7] The question of what is 'usual' for estate agents and auctioneers will be considered in some detail in later chapters. At this stage the point may be illustrated by examples from other areas.

---

1 See pp 236–237.
2 See *Papillon* v *Brunton* (1860) 5 H & N 518.
3 *Pearse* v *Bolter* (1860) 2 F & F 133.
4 See *Freeman & Lockyer (a firm)* v *Buckhurst Park Properties (Mangal) Ltd* [1964] 2 QB 480.
5 *Houghton & Co* v *Nothard, Lowe & Wills Ltd* [1927] 1 KB 246.
6 *Bayliffe* v *Butterworth* (1847) 1 Ex 425.
7 *Blackburn* v *Mason* (1893) 68 LT 510.

A broker selling goods, unlike an ordinary agent, is impliedly authorised to extend reasonable credit to a buyer.[1] No credit, however, may be given on a sale of shares, for this would not be according to the usual course of business.[2] An insurance broker, in the capacity of agent for an insurance company, has usual authority to enter into an interim contract of insurance, albeit orally and in informal or colloquial language.[3] Again, the managing agent of a block of flats has usual authority to make representations to a prospective tenant as to the condition of the property.[4]

As for the legal profession, it was held in *Domb* v *Isoz*[5] that a solicitor authorised to make a contract for the sale of land can effect that contract in any manner recognised by law (in the case itself, this was by an agreement over the telephone between the solicitors that contracts should be deemed to be exchanged). And in *Waugh* v *H B Clifford & Sons Ltd*[6] it was held that, while the implied authority of a solicitor or barrister to compromise a legal action may be limited *vis-à-vis* the client by such things as the agent's knowledge of the client's financial circumstances, this would not prevent the other party from treating any agreed compromise as valid; from the third party's point of view, the only limitation on the agent's usual authority is that the compromise agreement must not contain any matter which is 'collateral to the suit'.

The last example serves to emphasise the point that a principal seeking to restrict the usual authority of an agent must bring that restriction to the attention of the relevant party. Informing the agent will mean that, if the restriction is exceeded, the agent will be liable to the principal for disobedience. However, in order to prevent a third party from relying on the authority which an agent would normally possess, it is necessary for the principal to inform that third party that this particular agent's authority has been limited.

This focus on what the third party knows makes it clear that the underlying basis of usual authority is some form of estoppel; the principal, having produced the appearance of authority, is not permitted to deny that it actually exists. This makes usual authority

---

1   *Boorman* v *Brown* (1842) 3 QB 511.
2   *Wiltshire* v *Sims* (1808) 1 Camp 258.
3   *Stockton* v *Mason* [1979] RTR 130.
4   *Gordon* v *Selico Co Ltd* [1986] 1 EGLR 71.
5   [1980] Ch 548.
6   [1982] Ch 374.

very similar to apparent authority and, indeed, many cases could be decided equally well upon either basis. However, the two concepts are not entirely coextensive, because of a line of cases[1] in which it has been held that a third party may rely on the 'usual authority' of a person who is not known to be an agent. In such circumstances it would be completely unrealistic to suggest that the principal has given the agent any 'appearance of authority', and so an independent doctrine of usual authority must be recognised.[2]

Implication from necessity

In a limited number of cases, where a person's property or interests are threatened by some sort of emergency, another person may be legally justified in acting to protect them. The term 'agency of necessity' is often used to encompass these situations, although this is misleading, since it is extremely rare for the so-called 'agent' to create a binding relationship between the 'principal' and a third party.[3] More commonly, the point arises because the Good Samaritan in question demands reimbursement of whatever expenses he or she has incurred, whereupon it must be decided whether or not the intervention was justified.

It should be emphasised at the outset that English law has never looked with much favour upon the providers of unsolicited services, so that, for example, claims in respect of expenses incurred in looking after a stray dog,[4] or in removing timber found on a river bank to a place of safety,[5] have proved unsuccessful. However, the courts are rather more sympathetic where an existing agent oversteps the normal limits of authority in an emergency. Thus, for example, a railway company has been held bound by the actions of employees in ordering medical attention and food for injured passengers following an accident.[6] On

---

1   Of which the best known is *Watteau* v *Fenwick* [1893] 1 QB 346.

2   This doctrine is considered at p 12.

3   An example of the rare case is that of a shipmaster who may, in an emergency, sell both the ship and its cargo: *The Gratitudine* (1801) 3 Ch Rob 240; *Robertson* v *Caruthers* (1819) 2 Stark 571.

4   *Binstead* v *Buck* (1776) 2 W Bl 1117.

5   *Nicholson* v *Chapman* (1793) H Bl 254.

6   *Walker* v *Great Western Railway Co* (1867) LR 2 Ex 228; *Langan* v *Great Western Railway Co* (1873) 30 LT 173.

the other hand, where a mine manager borrowed money to pay arrears of miners' wages and thus prevent them from distraining upon the mine machinery, it was held that this could not be justified on the ground of necessity.[1]

Even in those situations where agency of necessity is capable of being invoked, it is subject to such strict conditions that attempts to rely on it rarely succeed. In the first place, the 'agent' must establish that it was impossible in practical terms to obtain instructions from the 'principal',[2] a requirement which modern communications have rendered difficult to satisfy. Secondly, there must be a real emergency, something which the courts have proved reluctant to recognise except in relation to livestock or perishable goods.[3] Finally, it should be noted that inconvenience to the 'agent' does not constitute 'necessity' for this purpose, so that a person who allowed the plaintiff to store furniture free of charge in her house, but who then sold it because she needed the space and could not contact the plaintiff, was held liable in the tort of conversion; she could not evade liability by claiming to have acted as an agent of necessity.[4]

Apparent authority

We have already referred more than once to the doctrine of 'apparent authority', under which a principal may be held responsible for what an agent does, even though the agent may have disobeyed explicit instructions. The legal rule which produces this conclusion has been expressed as follows:

> Where a person, by words or conduct, represents or permits it to be represented that another person has authority to act on his behalf, he is bound by the acts of that other person with respect to anyone dealing with him as an agent on the faith of any such representation, to the same extent as if such other person had the authority that he was represented to have, even though he had no such actual authority.[5]

---

1   *Hawtayne* v *Bourne* (1841) 7 M & W 595.

2   *Springer* v *Great Western Railway Co* [1921] 1 KB 257.

3   See *Prager* v *Blatspiel, Stamp & Heacock Ltd* [1924] 1 KB 566 (no agency of necessity to sell furs which could have been stored without risk of deterioration).

4   *Sachs* v *Miklos* [1948] 2 KB 23. See also *Munro* v *Willmott* [1949] 1 KB 295.

5   *Bowstead & Reynolds on Agency* (17th ed), 8–013.

Nature and effect of apparent authority

The above description of apparent authority makes clear that it is concerned exclusively with the relationship between principal and third party. It is a means by which the principal becomes liable to the third party as if the agent had been authorised, when in truth this was not so. The justification for imposing such liability upon the principal lies in the representation which he or she has made and on which the third party has relied. It is thus based on, or at least closely linked to, the legal doctrine of estoppel. An important conclusion to be drawn from this is that, where the representation of authority comes from the agent, the principal will seldom be held liable. It is inherently improbable that an agent, who has no general authority to carry out some act, is none the less authorised to confirm that he or she has been given special authority to carry it out.[1]

The simplest and most straightforward case of apparent authority (albeit one which is seldom encountered) arises out of an express statement made by the principal. In *Trickett* v *Tomlinson*,[2] for example, the principal told a third party that he had authorised his agent to settle a dispute between them, but did not mention that he had placed limits upon the agent's discretion. When the agent reached a settlement outside these limits, the principal was held bound by it.

More commonly, the 'representation' on which apparent authority rests is to be inferred from the principal's conduct (which may be purely passive). Thus, for example, a principal who permits an agent to order goods from a particular supplier, in the sense of meeting without demur a series of bills from that supplier, may well find that the agent is clothed with the appearance of authority to continue in this way.[3] In *Townsend Carriers Ltd* v *Pfizer Ltd*,[4] a notice to terminate a business tenancy was served on an associated company of the tenant by an associated company of the landlord. The service was held to be valid, since these associated companies had handled all the parties' dealings throughout the lease, and they had therefore been given the appearance of the necessary authority by their respective principals.

---

1   *Armagas Ltd* v *Mundogas SA* [1986] AC 717. Improbable, but not impossible: see *First Energy (UK) Ltd* v *Hungarian Bank Ltd* [1993] 2 Lloyd's Rep 194.
2   (1863) 13 CBNS 663.
3   *Summers* v *Solomon* (1857) 7 E & B 879.
4   (1977) 33 P & CR 361.

In practical terms, perhaps the most important way of 'holding out' an agent as having authority consists of appointing that agent to some position in the principal's business. Where this is done, third parties are entitled to assume, until they are notified to the contrary, that the agent is invested with the authority which is usual in that position and, further, that this authority has not been in any way restricted. In the case of a valid appointment, the principal's liability to the third party might equally be based upon the doctrine of implied or 'usual' authority.[1] However, the value of apparent authority is that it can be relied on by a third party even where the agent was never properly appointed at all.[2]

Assuming that a 'representation' as to an agent's authority has been made, a third party who seeks to hold the principal bound must show that he or she relied on it.[3] This of course means that apparent authority can never be used by a third party who knows that the agent has no real authority.[4] Less obviously, it has been held to mean that, where an agent acts under the authority of a written document, the third party is treated as being aware of any lack of authority which would be revealed by reading that document.[5] Moreover, where a transaction entered into by an agent is so 'irregular' as to arouse suspicion, a third party can no longer rely on apparent authority but must make enquiries.[6] Thus, for example, where a solicitor authorised to sign cheques for a client dishonestly used one of these cheques to buy a car for himself, it was held that the car dealers should have checked directly with the client to see that this purchase was authorised.[7]

Where apparent authority exists, it takes effect in a number of different ways. First (although this does not often occur), it may give an appearance of authority to someone who in truth has never been the principal's agent at all.[8] Second, it may serve as a continuation, for the

---

1   See p 6.
2   See *Freeman & Lockyer (a firm)* v *Buckhurst Park Properties (Mangal) Ltd* [1964] 2 QB 480.
3   *Farquharson Brothers & Co* v *King & Co* [1902] AC 325, 341.
4   As to the common practice of denying that estate agents or auctioneers are authorised to make statements about property for sale, see pp 213–215.
5   *Jacobs* v *Morris* [1902] 1 Ch 816.
6   *Lloyds Bank Ltd* v *Chartered Bank of India, Australia & China* [1929] 1 KB 40.
7   *Reckitt* v *Barnett, Pembroke & Slater Ltd* [1929] AC 176. See also *Midland Bank Ltd* v *Reckitt* [1933] AC 1.
8   For an example, see *Mildner (F) & Sons* v *Noble* (1956) *The Times*, March 8; [1956] CLY 32.

benefit of a third party, of an agency which as between principal and agent has ceased to exist.[1] Third, and most commonly, it may prevent a restriction upon the agent's express or implied authority from operating to the detriment of a third party who has no notice of it.[2]

Analogous forms of authority

Attention must now be given to two situations in which, although not all the elements of apparent authority are present, the courts have none the less held a principal responsible for what an agent has done. The first of these arose in the notorious and much-criticised case of *Watteau* v *Fenwick*,[3] which concerned a beer-house purchased by the defendant brewers from a Mr Humble. The defendants kept Humble on as manager of the establishment, and it was his name which appeared over the door as licensee; in consequence, third parties had no reason to suspect that he was no longer the owner of the business, but was merely agent for an undisclosed principal. The defendants instructed Humble not to buy supplies for the business but he, in contravention of this instruction, purchased cigars on credit from the plaintiff and failed to pay for them. The Divisional Court of Queen's Bench[4] held the defendants liable to the plaintiff for the price of these cigars, on the basis that:

> The principal is liable for all the acts of the agent which are within the authority usually confided to an agent of that character, notwithstanding limitations, as between the principal and the agent, put upon that authority'.

The decision in *Watteau* v *Fenwick* cannot be accommodated within the ordinary principles of apparent authority, because it can hardly be said that a person who does not appear to be an agent in the first place has been 'held out' as possessing any authority at all. Indeed, had the plaintiff been aware that Humble was merely an agent, the decision would probably have been different, since it was (and is) common for the manager of licensed premises to be restricted in the choice of suppliers.[5] The case has further been criticised on the ground that there

---

1   As in *Summers* v *Solomon* (1857) 7 E & B 879: p 10.
2   As in *Trickett* v *Tomlinson* (1863) 13 CBNS 663.
3   [1893] 1 QB 346.
4   Relying on *Edmunds* v *Bushell and Jones* (1865) LR 1 QB 97.
5   See *Daun* v *Simmins* (1879) 41 LT 783.

was little reason for allowing the plaintiff, who presumably intended to give credit exclusively to Humble, to hold the undisclosed principal responsible on a transaction which the latter had expressly forbidden. None the less, the decision has been followed on similar facts.[1]

The second judicial extension of apparent authority is based on the decision of the Court of Appeal in *Spiro v Lintern*.[2] The defendant there asked his wife to look for a purchaser for a house of which he was the sole owner. The wife, however, went further and, without disclosing that she was acting as an agent, purported to exchange contracts with the plaintiff for the sale of the property. When the existence of the defendant was revealed, he made no reference to his wife's lack of authority; indeed, he appeared to treat the contract as binding, for he allowed the plaintiff to send an architect and a builder to carry out alterations to the property. It was held that the defendant was bound by the contract; by failing to inform the plaintiff to the contrary, he was liable to the same extent as if his wife had been authorised.

Although this decision is clearly not a straightforward application of apparent authority (since there was no such 'appearance' when the contract was signed), it might be regarded as an example of ratification,[3] a doctrine by which a principal can give retrospective authority for the act of an agent. However, it is established that there can be no ratification by an undisclosed principal,[4] and so *Spiro v Lintern* cannot be justified on that ground. It is rather a case of the defendant being taken to represent that his wife had authority to make the contract, and then, once the plaintiff had acted upon this representation by incurring expense, being estopped from denying its truth.

The type of authority recognised in *Spiro v Lintern* was again utilised by the Court of Appeal in *Worboys v Carter*,[5] which arose when the defendant, a tenant farmer, was sent to prison. A land agent representing the defendant purported to arrange an assignment of his tenancy to the plaintiff. The defendant, although never intending to go through with this transaction, allowed the plaintiff to believe that it was valid and that he could safely proceed to sell his own farm. Again the defendant was held bound by his agent's unauthorised act; not on the

---

1   *Kinahan & Co Ltd v Parry* [1910] 2 KB 389; reversed on other grounds [1911] 1 KB 459.

2   [1973] 3 All ER 319.

3   See p 14.

4   *Keighley, Maxsted & Co v Durant* [1901] AC 240.

5   [1987] 2 EGLR 1.

basis of ratification (which the Court of Appeal thought inapplicable, given the clear evidence that the defendant had no intention of adopting the agent's acts), but rather on the simple ground of estoppel.

Ratification

In certain circumstances an agent's unauthorised act may be subsequently adopted by the agent's principal. The principal in such cases is said to 'ratify' the agent's act, and this operates, by and large, to give retrospective authority to the agent. The precise effects of this legal doctrine, and the conditions necessary for its operation, may now be considered.

Effects of ratification

As a general rule, ratification operates to place the principal, the agent and the third party in the positions in which they would have been had the agent acted with authority in the first place. If the agent has purported to make a contract on the principal's behalf, the effect of this will be back-dated to the time when it was made. Furthermore, since the agent is now deemed to have acted with authority, there can be no personal liability to either the principal or the third party. The only limitation appears to be that there is no implication of authority for the agent to act similarly in future.[1]

It cannot be denied that the doctrine of ratification is capable of producing some strange results, of which the strangest is perhaps the controversial case of *Bolton Partners* v *Lambert*.[2] The defendant there made an offer to purchase the plaintiffs' property which the plaintiffs' agent, without any authority, accepted. The defendant notified the plaintiffs that his offer was withdrawn, whereupon the plaintiffs replied that they were ratifying their agent's acceptance of it! The Court of Appeal held that the plaintiffs' ratification operated retrospectively to create a binding contract from the date of the agent's acceptance, and that the defendant's withdrawal was thus of no legal effect.

---

1   *Irvine* v *Union Bank of Australia* (1877) 2 App Cas 366.
2   (1889) 41 ChD 295; much criticised, but followed in *Re Tiedemann and Ledermann Frères* (1899) 2 QB 66.

The difficulty which results from *Bolton Partners* v *Lambert* is that the third party is placed entirely at the mercy of the principal, who is free to decide whether or not to ratify. The third party can therefore neither ensure that there is a binding contract, nor prevent one from coming into operation. Perhaps in recognition of the hardship which may thus arise, the courts have kept the 'retroactivity' principle within fairly strict limits. It has, for example, been held that, where an agent's action is expressly[1] or impliedly[2] made 'subject to ratification', it will take effect only when the principal actually ratifies it. Furthermore, if the agent actually agrees with the third party to cancel a contract which they have purported to make, the principal cannot then revive it by ratification.[3]

Conditions for ratification

The first requirement of a valid ratification is that the principal must exist as a legal person and have the necessary capacity to act, not only at the time of the ratification, but also at the time of the agent's action. It is because of this that a limited company cannot ratify anything done on its behalf before its incorporation.[4] Thus, where a survey of a flat was carried out on behalf of a company which was to be formed to purchase it, the company could not ratify its agent's act so as to sue the surveyor for beach of contract; however, it was held that the surveyor owed the unformed company a duty of care in tort, and could be liable for a breach of that duty![5]

Ratification can only be carried out by the person on whose behalf the agent has purported to act.[6] The most important application of this rule is that there can be no ratification by a principal who, at the time of the agent's act, was undisclosed.[7] The leading case on this point is *Keighley, Maxsted & Co* v *Durant*,[8] where the defendants authorised an

---

1   *Watson* v *Davies* [1931] 1 Ch 455.
2   *Warehousing & Forwarding Co of East Africa Ltd* v *Jafferali & Sons Ltd* [1964] AC 1.
3   *Walter* v *James* (1871) LR 6 Ex 124.
4   *Kelner* v *Baxter* (1866) LR 2 CP 174. The 'agent' in such a case is personally liable to the third party: Companies Act 1985, s 36C(1).
5   *Miro Properties Ltd* v *J Trevor & Sons* [1989] 1 EGLR 151.
6   *Saunderson* v *Griffiths* (1826) 5 B & C 909.
7   There are even suggestions that the principal must have been identified: see *Watson* v *Swann* (1862) 11 CBNS 756.
8   [1901] AC 240.

agent to buy wheat on a joint account for himself and them. The agent bought wheat from the plaintiff without disclosing the joint account, at a price in excess of that which the defendants had authorised. The defendants accepted what their agent had done but, when the plaintiff sued upon the contract, the House of Lords held that the defendants, as undisclosed principals, could not be said to have ratified and were therefore not liable.[1]

Unless the principal is seeking to ratify the agent's execution of a deed (in which case the ratification too must be by deed),[2] there are in general no particular formalities to be observed. Indeed, many cases of ratification are based on some conduct of the principal, as where the principal sells goods which have been purchased by an agent in excess of authority.[3] In accordance with this principle, it has been held that a principal who adopts an unauthorised sale of property will normally also be taken to ratify the appointment of an agent to sell it, so as to incur liability for that agent's commission.[4] However, a court will not infer ratification from the principal's conduct where the principal had no choice whether to accept or reject what the agent had done,[5] nor where the principal did not know all the material facts about the agent's unauthorised act.[6]

In an effort to protect third parties against the hardship which may be caused by the retrospective nature of ratification, the courts have imposed a number of other restrictions upon the use of this doctrine. In the first place, it is settled that, where an act has to be performed within a certain time, it must, if performed without authority, be ratified within that time.[7] Thus, where a landlord's agent serves a notice to quit on a tenant without authority, this may not be ratified after the last date on which the notice could validly have been served.[8] If no time is fixed, ratification must take place within a reasonable

---

1   Note the possibility of estoppel in such circumstances: *Spiro v Lintern* [1973] 3 All ER 319; p 13.
2   *Hunter v Parker* (1840) 7 M & W 322; *Oxford Corp v Crow* [1893] 3 Ch 535.
3   *Cornwal v Wilson* (1750) 1 Ves 509.
4   *Keay v Fenwick* (1876) 1 CPD 745; *cf Hughes v Hughes* (1971) 221 EG 145.
5   *Forman & Co Proprietary Ltd v The Liddesdale* [1900] AC 190.
6   *The Bonita, The Charlotte* (1861) 1 Lush 252.
7   But an unauthorised issue of legal proceedings by an agent may be ratified, even after the limitation period has expired: *Presentaciones Musicales SA v Secunda* [1994] Ch 271.
8   *Doe d Mann v Walters* (1830) 10 B & C 626.

time.[1] Second, ratification will not be allowed where its effect would be to divest third parties of proprietary rights which have already vested in them.[2] Finally, where a third party does some act which is lawful when done (such as refusing to hand over the principal's goods on demand by an unauthorised agent), subsequent ratification of the agent's act cannot retrospectively make the third party's act unlawful.[3]

Agency transactions

Having examined the various types of authority which an agent may possess, we may now consider in rather more detail the ways in which a principal may be held legally responsible for what the agent does. The most important area concerns the making of contracts, but other matters are also relevant.

Contracts

The extent to which the principal is bound by contracts entered into by an agent may vary according to whether the third party was aware of the principal's existence. Moreover, a third party who has some defence against the agent may be able to set up that defence against the principal; and the principal may in some circumstances be able to do likewise against the third party.

Disclosed and undisclosed principals

The legal difference between a disclosed and an undisclosed principal turns, not on whether the third party knew the principal's identity, but rather on whether the third party was aware of the principal's existence. Where the agent is known to be acting as an agent (as will be almost always be the case with an estate agent or auctioneer), the vendor will be a disclosed (though quite possibly unnamed) principal. The true undisclosed principal is found in those cases where the agent

---

1   *Metropolitan Asylums Board* v *Kingham & Sons* (1890) 6 TLR 217.

2   *Bird* v *Brown* (1850) 4 Exch 786; *Dibbins* v *Dibbins* [1896] 2 Ch 348; *Presentaciones Musicales SA* v *Secunda* [1994] Ch 271; *Smith* v *Henniker-Major & Co* [2003] Ch 182; *The Borvigilant* [2003] 2 Lloyd's Rep 520.

3   *Solomons* v *Dawes* (1794) 1 Esp 83.

appears to the third party to be contracting personally, and where the principal only comes into the picture at a later stage.

Where an agent, acting within the scope of his or her authority,[1] enters into a contract on behalf of a disclosed principal, the general rule (subject to a few minor exceptions) is that the latter is both liable and entitled on that contract. This is central to the law of agency; indeed it is the fundamental purpose for which agency was designed. The only qualifications relate to contracts made in a particular form, most notably by deed.[2] At common law, a principal could only be a party to such a contract if the deed was executed by the agent in the principal's name,[3] or if the agent could be regarded as a trustee of the principal's rights.[4] This position has been modified by statute;[5] however, it appears that the principal must still be named as a party to the deed in order to acquire any rights or liabilities under it.

The position of a disclosed principal is just as one would expect. What may come as more of a surprise, however, is the fact that an undisclosed principal is to a large extent in exactly the same situation. Notwithstanding that the third party has acted throughout on the assumption that the 'agent' is dealing personally, the principal becomes both liable and entitled on the contract made.[6] Of course, this will depend upon the agent having acted with authority, and it should be remembered that there can be no question of ratification,[7] although the principal may be estopped by subsequent conduct from denying the agent's authority.[8] Further, the doctrine of apparent authority is not relevant, since the principal cannot be said to 'hold out' the agent. However, the principal will be bound by acts falling within the agent's 'usual authority', even where these have been expressly forbidden.[9]

The doctrine of the undisclosed principal appears somewhat anomalous, and the courts have attempted to ensure that it does not

---

1 Including *any* of the types of authority described above.
2 There are also special rules governing bills of exchange, cheques and promissory notes: see the Bills of Exchange Act 1882, ss 17, 23 and 91.
3 *Schack* v *Antony* (1813) 1 M & S 573.
4 *Harmer* v *Armstrong* [1934] Ch 65.
5 Powers of Attorney Act 1971, s 7(1), replacing the Law of Property Act 1925, s 123(1).
6 The agent too remains a party to the contract: *Saxon* v *Blake* (1861) 29 Beav 438.
7 *Keighley, Maxsted & Co* v *Durant* [1901] AC 240.
8 *Spiro* v *Lintern* [1973] 3 All ER 319: see p 13.
9 *Watteau* v *Fenwick* [1893] 1 QB 346: p 12.

operate harshly against third parties. In general, it appears that an undisclosed principal will not be allowed to intervene where this would materially weaken the position of the third party. In *Hanstown Properties Ltd* v *Green*,[1] for example, where two tenants of a flat claimed that they had entered into the lease as agents for themselves and a third person. To allow this third person to come into the picture would be highly detrimental to the landlords, since it would increase the number of persons who might qualify for a statutory tenancy. Accordingly, the undisclosed principal was not permitted to reveal herself as an additional tenant.

In addition to this general regard for the interests of the third party, the courts have on occasion held that the contractual description of one of the parties is such as to exclude the possibility that that party is acting as an agent, and thus to rule out any undisclosed principal.[2] It should be acknowledged, however, that this approach is somewhat controversial, and will in any event not apply to more equivocal descriptions of a party, such as 'tenant' or 'landlord'.[3]

It might be thought that, in any case where the third party can show that the agent's identity is a matter of concern, this would be sufficient to keep an undisclosed principal out of the picture. The reality, however, is not quite so simple. It is certainly true that, where the law regards a contract as too 'personal' to be assigned (which will be the case, for example, where it involves particular skills or characteristics of a party), it will equally not be possible for an undisclosed principal to take over that contract. However, it appears that the third party may not exclude an undisclosed principal merely by showing that he or she would not have been willing to contract with that person. Thus, surprising though it might seem, a person who knows that he or she would not be acceptable to another as a contracting partner may evade that restriction by operating as an undisclosed principal.

This last point appears from the case of *Dyster* v *Randall & Sons*,[4] where the plaintiff, who had been employed by the defendants, was dismissed by them in circumstances which left no doubt that they would refuse to do business of any kind with him. Wishing to obtain

---

1   [1978] 1 EGLR 85.
2   Eg where the agent is described as 'owner' (*Humble* v *Hunter* (1848) 12 QB 310) or 'proprietor' (*Formby Bros* v *Formby* (1910) 102 LT 116).
3   See *Danziger* v *Thompson* [1944] KB 654; *Epps* v *Rothnie* [1945] KB 562.
4   [1926] Ch 932.

two plots of land from the defendants, the plaintiff persuaded a friend to buy them as his agent. When the defendants, on learning the truth, purported to cancel the contract, the plaintiff sued for and obtained a decree of specific performance. The court held that there had been no positive misrepresentation as to the purchaser's identity,[1] and that this was not a material ingredient of a contract for the sale of land.[2]

Principal's defences

A principal (whether disclosed or undisclosed) who is apparently bound by a contract made by an agent may nevertheless be able to raise one or more defences when sued by the third party. In the first place, it is an established rule of common law that a third party who has elected to give credit exclusively to one person may not subsequently choose to proceed against the other. What constitutes an 'election' for this purpose is a question of fact, but it may be stated that the courts require evidence that the third party has unequivocally decided to look exclusively for payment to the agent;[3] thus, merely receiving partial payment from the agent,[4] debiting the agent in accounts[5] or even commencing legal proceedings[6] may all fall short of what is required to release the principal. In any event, there can be no election by someone who does not know all the material facts, which means that the doctrine cannot apply to things done by the third party while the principal remains undisclosed.

A question which sometimes arises is whether a principal who owes money to the third party may discharge that obligation by paying the agent. At one time this was held to be so, on the general ground that it would be unfair to make the principal pay twice.[7] However, it now appears that the principal may only rely upon a payment to the agent as discharging an obligation to the third party in cases where the principal has been induced by the third party to make the payment (for

---

1   Had there been, the contract would not have been enforceable: *Archer* v *Stone* (1898) 78 LT 34.
2   For a contract in which identity was held to be material, see *Said* v *Butt* [1920] 3 KB 497.
3   See, for example, *Chestertons* v *Barone* [1987] 1 EGLR 15.
4   *Ex p Pitt* (1923) 40 TLR 5.
5   *Young & Co Ltd* v *White* (1911) 28 TLR 87.
6   *Clarkson Booker Ltd* v *Andjel* [1964] 2 QB 775.
7   *Armstrong* v *Stokes* (1872) LR 7 QB 598.

example where the third party leads the principal to believe that the agent has already paid the debt).[1] An important consequence is that an undisclosed principal can never avoid liability on this ground, since a third party can hardly be said to induce payment by making representations to a person whose very existence is unknown.[2]

Third party's defences

The converse situation to that described above is where a third party who owes money to the principal claims to have discharged that debt by paying the agent, or by setting off a sum which the agent owes to the third party. The extent to which such claims will succeed depends, among other things, on whether the principal is disclosed or undisclosed.

In the case of a disclosed principal, everything turns on the scope of the agent's authority to receive payment. This is likely to be limited to the receipt of cash, in which case payment in kind will not discharge the debt.[3] Similarly, it is not often that an agent will have authority to set off a personal debt in partial satisfaction of what the third party owes to the principal, although this has occasionally been established.[4]

Where the principal is undisclosed, it might be thought that any payment or set-off by the third party before learning of the principal's existence would validly discharge the debt. This is largely correct, although it is subject to a rather illogical qualification as a result of the decision in *Cooke & Sons* v *Eshelby*.[5] It was held by the House of Lords in that case that, unless the principal has in some way induced the third party to believe that the agent is contracting personally (ie has enabled the agent to appear to deal as principal), the principal will not be bound by any settlement or set-off. While this qualification might seem somewhat unfair, it will not often create hardship for the third party, as the principal will be held to have enabled the agent to appear

---

1   *Wyatt* v *Marquis of Hertford* (1802) 3 East 147.
2   *Irvine & Co* v *Watson and Sons* (1880) 5 QBD 414, approving *Heald* v *Kenworthy* (1855) 10 Exch 739.
3   *Howard* v *Chapman* (1831) 4 C & P 508 (a quantity of horse-hair). More importantly, nor will a bill of exchange or cheque suffice: *Williams* v *Evans* (1866) LR 1 QB 352.
4   See, for example, *Barker* v *Greenwood* (1837) 2 Y & C Ex 414.
5   (1887) 12 App Cas 271.

as principal in any case where the agent is entrusted with actual possession of goods to sell.[1]

As to other defences which the third party may seek to set up, it is clear that a principal who is personally guilty of fraud or misrepresentation runs the risk of losing the contract or having to pay damages. In *Ludgater* v *Love*,[2] for example, the defendant had some sheep for sale which he knew to be diseased. He appointed an agent to sell the sheep, but concealed their condition from him. The agent sold the sheep as sound, which was what the defendant had intended, and the purchaser was held entitled to damages from the defendant for this fraud.

Where the third party's defence against contractual liability (for example on the ground of fraud, misrepresentation or illegality) is based on some act of the agent, the legal position depends on whether the principal is disclosed or undisclosed. A disclosed principal is bound by whatever the agent is authorised to do, so that the third party is placed in the same position as if the principal had done the act in person. Thus in *Mullens* v *Miller*,[3] an estate agent's misrepresentation about the property prevented the client from obtaining specific performance of the contract of sale; in *Gosling* v *Anderson*,[4] on similar facts, a purchaser recovered damages from the vendor. So too, in *Refuge Assurance Co Ltd* v *Kettlewell*[5] the holder of an insurance policy was held entitled to recover from the company premiums which she had paid on the faith of false statements made by the company's agents. Finally, in *Biggs* v *Lawrence*[6] where one of three partners sold goods which he knew to have been packed for the purpose of smuggling, it was held that the firm was prevented by this illegality from suing the purchaser for the price of the goods.

Where the principal is undisclosed, it appears that the third party is in a stronger position for, in addition to the rights outlined above, the third party may rely upon any defences which are obtained against the agent before the principal's existence becomes known.[7]

---

1   *Borries* v *Imperial Ottoman Bank* (1873) LR 9 CP 38.
2   (1881) 44 LT 694.
3   (1882) 22 ChD 194.
4   (1972) 223 EG 1743.
5   [1909] AC 243.
6   (1789) 3 TR 454.
7   At least where the principal has enabled the agent to appear to deal in person: see *Cooke & Sons* v *Eshelby* (1887) 12 App Cas 271; p 21.

Torts

It may be stated at the outset that a person who expressly authorises or subsequently ratifies the commission of a tortious act by an agent will incur liability in respect of it. Such liability is regarded as 'personal', in the sense that the elements of whatever tort is in issue (particularly in relation to knowledge or state of mind) must be proved against the principal.

Leaving aside cases of this kind, a principal is not generally responsible for an agent's tort, unless the principal and agent in question are also employer and employee. However, cases of misrepresentation form an important exception to this general principle.

Actions for deceit and negligent misstatement are undoubtedly tortious in nature. Nevertheless, the statements which give rise to such actions are frequently made by agents who are employed either to make a contract or at least to negotiate one. Perhaps because of this interweaving of the rules of tort and contract, it is well established that a principal is liable for fraudulent[1] statements made by the agent within the scope of authority (which, it should be remembered, includes implied usual authority).[2] Thus in *Gordon* v *Selico Co Ltd*,[3] where a builder deliberately covered up dry rot in a flat which was about to be let, it was held that the landlords could not be held directly responsible for this, since it is not within a builder's authority to make statements about a property.[4] However, the evidence showed that the managing agents of this block of flats knew what the builder had done and were thus also guilty of fraud; the landlords were therefore responsible, since the making of statements about a property undoubtedly falls within the usual authority of managing agents.

The liability which is thus imposed upon the principal may arise even where the fraudulent statement is made for the agent's own purposes. In *Lloyd* v *Grace, Smith & Co*[5] the plaintiff sought advice from the defendant solicitors about two properties which she owned. She dealt entirely with the defendants' managing clerk, who fraudulently induced her to transfer the properties to him; he then disposed of them

---

1   And presumably negligent.
2   See *Mullens* v *Miller* (1882) 22 ChD 194: p 206.
3   [1986] 1 EGLR 71.
4   Deliberate concealment of defects is regarded in law as a 'statement' that none exist.
5   [1912] AC 716.

and misappropriated the proceeds of sale. The House of Lords held that, by allowing the managing clerk to deal with clients, the defendants had given him apparent authority to conduct certain types of business. The fraud which was committed fell within the scope of this apparent authority and the defendants were accordingly liable for it, notwithstanding that the clerk had acted entirely for his own benefit.

The tort of deceit, which consists of knowingly making a false statement with the intention that it should be acted upon, raises one particular problem in relation to agency. This is where the agent makes a false statement innocently, while the principal, who knows the truth, is unaware that the statement is being made.[1] In *Cornfoot* v *Fowke*,[2] an agent told the prospective tenant of a house that there was nothing objectionable about it. In truth, as the principal (though not the agent) knew all too well, the house was next door to a brothel. The tenant sued the principal for fraud, but the action failed. A similar point arose in the case of *Armstrong* v *Strain*,[3] which concerned the sale of a much-underpinned bungalow. The vendor did not mention this fact to the estate agents, who innocently misrepresented its condition to the purchaser. Devlin J, whose decision was affirmed by the Court of Appeal, concluded that neither vendor nor estate agent was liable for fraud, since 'you cannot add an innocent state of mind to an innocent state of mind and get as a result a dishonest state of mind'.[4]

Notice

A point of general commercial importance concerns the effect of serving notice on[5] an agent under some statute, rule of common law or contractual provision. Whether this constitutes a valid service on the principal depends ultimately on the extent of the agent's authority to receive the notice, and it may be noted that the usual authority of an estate agent or surveyor in this respect is extremely limited,[6] though

---

1   If the principal intends that the agent should make the offending statement, he is personally guilty of fraud: *Ludgater* v *Love* (1881) 44 LT 694.
2   (1840) 6 M & W 358.
3   [1952] 1 KB 232.
4   [1951] 1 TLR 856, 872.
5   Or, occasionally, by an agent: see *Lawrence Chemical Co Ltd* v *Rubenstein* [1982] 1 All ER 653.
6   See *Robert Baxendale Ltd* v *Davstone (Holdings) Ltd* [1982] 3 All ER 496.

that of a managing agent may be more extensive.[1] More surprisingly, perhaps, there is no general assumption that a solicitor has authority to receive notice on behalf of a client,[2] although evidence that such authority has in fact been given may not be too hard to find.[3]

Agent and principal

Wherever an agency relationship exists, the agent and the principal are subject to a range of mutual rights and obligations. The precise nature of these depends primarily upon the terms of the agency contract; where that contract is silent, terms will be implied by the general law. Indeed, even where the agency relationship is not based upon contract at all (a rare occurrence for professional agents such as estate agents and auctioneers), many of the agent's normal duties[4] and rights[5] will still apply.

A detailed examination of the rights and duties of estate agents and auctioneers will be found in the following two chapters.

Agent and third parties

In considering the legal position of an agent *vis-à-vis* the third parties with whom the agent deals on the principal's behalf, it should be made clear from the outset that an agent who is guilty of committing a wrong, such as a tort or a breach of trust, will be personally liable for this. It may be that the third party can hold the principal responsible for what the agent has done but, if so, this is an additional remedy for the third party which in no way exonerates the agent as the primary wrongdoer.

As for contracts entered into by agents, the legal position is not quite so clear. In some circumstances the agent will drop out of the picture altogether, so that the resulting contract concerns only the principal and the third party. In other situations, however, the agent

---

1   See *Peel Developments (South) Ltd* v *Siemens plc* [1992] 2 EGLR 85.
2   *Saffron Walden Second Benefit Building Society* v *Rayner* (1880) 14 ChD 406; *Tate* v *Hyslop* (1885) 15 QBD 368; *Singer* v *Trustee of the Property of Munro* [1981] 3 All ER 215.
3   See, for example, *Westway Homes Ltd* v *Moores* [1991] 2 EGLR 193.
4   The agent will, however, be under no positive duty to act on the principal's behalf, nor to obey with any instructions subsequently issued.
5   The agent will be entitled to reimbursement of expenses and indemnity against losses, but not to any remuneration as such.

will be regarded as a party to the contract, with rights against and obligations towards the third party. It is this problem area which now falls to be discussed, together with the question of an agent's liability to the third party for acting without authority from the principal.

Contractual relationships

Where an agent makes a contract without revealing the existence of his or her principal, the agent is regarded in law as contracting personally.[1] The third party, after all, has intended to contract with the agent, and this contractual link is not to be taken away, even though the undisclosed principal may also be liable and entitled on the contract. As to the agent's right to enforce the contract in such a situation, however, this is subject to any defence (such as fraud) which the third party could have raised against the principal. This might not appear entirely logical (since the third party by definition intended to contract exclusively with the agent), but it was approved in *Garnac Grain Co Inc* v *HMF Faure & Fairclough Ltd*[2] on the ground that, if this were not so, 'the fraudulent principal would indirectly be entitled to enjoy the fruits of his own fraud'.

The question whether or not the agent of a disclosed principal is to be treated as contracting personally turns, at least in theory, on the intention of the parties in the individual case. However, the reported cases in which this has been held to be so appear on examination to fall into a number of recognisable categories, which may now be described. It should be noted that, in all these cases, a third party sued by the agent may set up any defence (such as a set off) which could have been used against the principal.[3]

Form of contract

At common law, a principal could not be a party to a formal contract (ie one made by deed) which was executed in the agent's name or with the agent's seal. As a result, the agent in such cases was personally

---

1   *Sims* v *Bond* (1853) 5 B & Ad 389; *Saxon* v *Blake* (1861) 29 Beav 438.
2   [1966] 1 QB 650, 656.
3   *Atkinson* v *Cotesworth* (1825) 3 B & C 647.

liable.[1] The primary rule has now been altered by statute,[2] so as to enable the principal to be regarded as a party; however, whether this means that the agent is no longer liable or entitled has been left unclear.

As for contracts which are made in writing, the law is not so strict as in the case of deeds, but it none the less creates a serious trap for the unwary. The common law rule[3] here is that a person who signs a contract in his or her own name will be liable[4] and entitled[5] upon it, even though it is understood on all sides that that person is merely acting as an agent. To avoid this conclusion, the contract documents themselves must reveal the agent's true status, for example by some qualification to the signature itself.

As to what kind of 'qualification' will release the agent, the case law has drawn a crucial distinction between words which clearly show the representative nature of the signature, and those which merely describe the position held by the signatory. Such epithets as 'Chartered Civil Engineer[6] or 'director'[7] have been treated as mere description. Nor can the addition of 'broker'[8] or 'agent'[9] be regarded as conclusive, since these may not show unequivocally that the person is signing the contract in that capacity. It seems that what is required is a phrase equivalent to 'for and on behalf of'[10] or 'on account of',[11] although it should be noted that even the last of these has not always proved infallible in excluding an agent's personal liability.[12]

Whether such formalistic rules are really appropriate to modern business conditions must be open to doubt, and it is worth pointing out that the Court of Appeal has refused to extend them to oral contracts. In such cases it is for the third party to show that the agent

---

1   *Appleton* v *Binks* (1804) 5 East 148.
2   Powers of Attorney Act 1971, s 7, replacing the Law of Property Act 1925, s 123(1).
3   Which is given statutory recognition, in relation to negotiable instruments, by the Bills of Exchange Act 1882, s 26(1).
4   *Higgins* v *Senior* (1841) 8 M & W 834.
5   *Short* v *Spackman* (1831) 2 B & Ad 962.
6   *Sika Contracts Ltd* v *BL Gill and Closeglen Properties Ltd* (1978) 9 Build LR 15.
7   *McCollin* v *Gilpin* (1881) 6 QBD 516.
8   *Hutcheson & Co* v *Eaton & Son* (1884) 13 QBD 861.
9   *Parker* v *Winlow* (1857) 7 E & B 942.
10   *Universal Steam Navigation Co Ltd* v *James McKelvie & Co* [1923] AC 492.
11   *Fairlie* v *Fenton* (1870) LR 5 Ex 169; *Gadd* v *Houghton* (1876) 1 ExD 357.
12   *Punjab National Bank* v *de Bonville* [1992] 3 All ER 104.

undertook personal responsibility; there is no presumption to this effect merely because the agent made the contract without using the magic words 'as agent'.[1]

Non-existent principal

Where a person claims to be acting as an agent for someone else, whether named or unnamed, this claim may turn out to have no foundation, either because the 'principal' has no existence at law (eg where it is an unformed company or an unincorporated association), or because the 'agent' really intends to deal personally but wishes to conceal this fact. In both cases, the law is usually prepared to treat the agent as personally liable and entitled on the contract. However whether this is a justifiable approach is a matter of considerable dispute.

Dealing first with the situation of a principal who lacks legal capacity, the presumption in such cases is that the third party intends to contract with the agent, since there would otherwise be no enforceable contract at all. Notwithstanding this possibility, however, the presumption is not conclusive. Thus in *Jones* v *Hope*,[2] where a solicitor rendered professional services to a volunteer corps (which had no legal personality of its own and could not therefore make a contract), it was held that the solicitor's fees could not be claimed from the commanding officer, as the evidence showed clearly that there had been no intention to make a contract with the colonel personally.

The most common example of this type of case concerns contracts made on behalf of a company which has not yet been formed. The position in such cases is now governed by the Companies Act. This provides that, subject to any agreement to the contrary, the contract shall take effect as one made personally with the 'agent', who is accordingly liable on it.[3]

The second situation mentioned above, that of an 'agent' who secretly intends to deal personally, is one in which the courts' decisions have been much more controversial. In *Schmaltz* v *Avery*[4] the plaintiff, describing himself as 'agent of the freighter', signed a charterparty

---

1   *N & J Vlassopulos Ltd* v *Ney Shipping Ltd, The Santa Carina* [1977] 1 Lloyd's Rep 478.
2   (1880) 3 TLR 247n.
3   Companies Act 1985, s 36C: see *Phonogram Ltd* v *Lane* [1982] QB 938. The agent is also entitled to enforce the contract: *Braymist Ltd* v *Wise Finance Co Ltd* [2002] 2 All ER 333.
4   (1851) 16 QB 655.

which provided; 'This charter being concluded on behalf of another party, it is agreed that all responsibility on the part of the agents shall cease as soon as the cargo is shipped.' Notwithstanding this provision, it was held that the plaintiff, on later revealing that he himself was the true principal, could enforce the contract in his own name.

In reaching this decision, the Court of Queen's Bench clearly believed that the identity of the principal must have been a matter of indifference to the third party, who had not bothered to ask who it was. The court acknowledged that, had the principal's identity been important to the third party, the decision might have been different. However, even this qualification failed to protect the third party in *Harper & Co v Vigers Bros*,[1] where the principle of *Schmaltz v Avery* was followed and indeed extended. The plaintiffs in the later case signed a charterparty 'by authority of and as agents for owners', at a time when they had no principal, but were speculating in freight and also wanted to obtain brokers' fees. The defendants gave evidence that, had they known that the plaintiffs had not already arranged a ship, they would not have made the contract. Notwithstanding this evidence, it was held that the plaintiffs were entitled to enforce the contract in their own name, although their concealment of the truth was enough to deprive them of an award of their legal costs.

It is perhaps possible to support these decisions on the basis that a third party who is prepared to contract with an unnamed principal cannot be unduly concerned about that principal's identity.[2] However, the same surely cannot be said of the situation where an agent actually names a principal before revealing that the latter does not exist. In such a case it might be assumed that the agent's misrepresentation would render the contract unenforceable, and yet, surprisingly, the law does not appear to adopt this position. Despite a strong dictum to the effect that 'where a man assigns to himself the character of agent to another whom he names, I am not aware that the law will permit him to shift his situation, and to declare himself the principal, and the other to be a mere creature of straw',[3] an action by such an agent to enforce a contract was successful in *Rayner v Grote*.[4]

---

1   [1909] 2 KB 549.
2   They have none the less been heavily criticised in the Scottish courts: see *Hills SS Co Ltd v Hugo Stinnes Ltd* [1941] SC 324, 340.
3   *Bickerton v Burrell* (1816) 5 M & S 383, 386.
4   (1846) 15 M & W 359.

An unexpressed reason for these decisions may have been the courts' desire to ensure that the 'agent' should not escape liability on the resulting contract, if the third party wishes to enforce it.[1] However, it should be pointed out that, if that is indeed what the courts wished to achieve, the agent in all such cases could have been held liable instead for breach of warranty of authority.[2]

Other cases

Although an agent is not automatically to be treated as contracting personally wherever the principal's identity is not revealed, there is no doubt that the courts are often influenced by this in their decision as to what the parties must have intended. Indeed, there are signs of an increasing readiness in the courts to conclude that, quite apart from the categories of case discussed above, a contract may well be entered into on the basis that the principal and the agent both undertake to guarantee performance.

A good example of this more flexible approach is provided by *The Swan*,[3] where the defendant formed a company to operate his fishing boat. The company, which hired the boat from the defendant in return for 90 per cent of its profits, was responsible for any repairs which might become necessary. Various repairs were carried out by the plaintiffs at the request of the defendant, and paid for by cheques drawn on the company's account. Eventually extensive repairs and alterations became necessary, and these were ordered by the defendant (signing himself 'director') on company notepaper, after a meeting between the defendant and the plaintiffs at which the work required was agreed. When the company became insolvent, the plaintiffs sued the defendant for the price of the repairs, alleging that he had made the contract either as principal or as agent in such circumstances that he was personally liable for its due performance. Brandon J while rejecting the plaintiffs' first contention, had little hesitation in accepting the second. It was, he said, natural to assume that a shipowner would undertake personal responsibility to pay for repairs. He might disown

---

1 Liability is readily imposed on the 'agent' in such circumstances: see, for example, *Hersom* v *Bernett* [1955] 1 QB 98; *Savills* v *Scott* [1988] 1 EGLR 20.

2 See p 31.

3 [1968] 1 Lloyd's Rep 5.

that responsibility, but only in clear terms; merely pointing out that he was placing the order on behalf of a company did not in itself relieve him of liability.

Liability for unauthorised acts

Where an agent purports to act on behalf of a principal, but exceeds whatever authority has been given, it follows that the principal will not be bound to the third party. In such circumstances, the agent may be personally liable to the third party, not on the contract itself (since the third party never intended to contract with the agent)[1] but rather for breach of an implied warranty that the agent had the necessary authority. In *Collen* v *Wright*,[2] the defendant land agent agreed on behalf of his principal to grant a 12 and a half-year lease of a farm to the plaintiff. The plaintiff brought an action against the principal to enforce this agreement, only to discover that the agent was not authorised to grant leases of this length. Accordingly a second action was brought against the defendant, who was held liable to pay damages. Willes J, delivering the judgment of a majority of the Exchequer Chamber, said:[3]

> A person, professing to contract as agent for another, impliedly, if not expressly, undertakes to or promises the person who enters into such contract, upon the faith of the professed agent being duly authorised, that the authority which he professes to have does in point of fact exist.

The liability of an agent in a case of this kind does not depend upon fraud[4] or even negligence, but rather on the fact that 'by professing to act as agent he impliedly contracts that he has authority, and it is immaterial whether he knew of the defect of his authority or not.'[5] Thus in *Yonge* v *Toynbee*,[6] where a solicitor continued to defend an action on behalf of a client who had been certified as insane, the Court of Appeal held the solicitor personally liable for the costs incurred by

---

1   This particular fallacy was exposed in *Lewis* v *Nicholson* (1852) 18 QB 503.
2   (1857) 8 E & B 647.
3   At p 657.
4   If the agent is aware of the lack of authority, an action will lie in the tort of deceit: *Polhill* v *Walter* (1832) 3 B & Ad 114.
5   *Yonge* v *Toynbee* [1910] 1 KB 215, 227, *per* Buckley LJ.
6   [1910] 1 KB 215.

the other party, notwithstanding that he neither knew of his client's insanity nor could reasonably have been expected to know of it.

The essence of this form of action lies in the third party's reliance on a representation that the agent possesses the necessary authority.[1] If there is no such representation (for example because the third party is informed as to any limitations on the agent's authority) then the agent cannot be liable. Thus in *Lily, Wilson & Co v Smales, Eccles & Co*,[2] where a firm of shipbrokers signed a contract 'by telegraphic authority', they were held not liable when it transpired that the telegram was inaccurate and that they were not in fact authorised to sign. Nor can liability arise where the 'representation' is one of law rather than of fact, since the third party is presumed to know what the law is and thus not to have been misled.[3]

The third party's 'reliance' in these cases is normally shown by entering into a contract, but this is not the only situation in which liability can arise. In *Starkey v Bank of England*,[4] for example, the Bank of England were held to have relied on a stockbroker's power of attorney (which turned out to have been forged) by transferring stock to a purchaser. Again, in *V/O Rasnoimport v Guthrie & Co Ltd*,[5] it was held that the agents of a shipowner who signed a bill of lading, without authority to do so, could be held liable to anyone who subsequently acquired that bill; the mere acquisition of it in the course of business, in the belief that it conferred rights which could be exercised against the shipowners, constituted sufficient 'reliance' on the agents' claim of authority to sign it.

The remedy in an action for breach of warranty of authority is an award of damages, assessed with the intention of putting the third party in the position which he or she would have occupied if the agent had been authorised and the principal had therefore been bound. In effect, therefore, the agent is forced to compensate the third party for whatever profitable bargain appeared to have been struck. Conversely, if the third party's rights against the principal would in any event have been worthless (for example because the principal is insolvent), then the damages awarded against the agent should be nominal.[6]

---

1   *Halbot v Lens* [1901] 1 Ch 344.
2   [1892] 1 QB 456.
3   *Eaglesfield v Marquis of Londonderry* (1875) 4 ChD 693; affirmed (1878) 38 LT 303.
4   [1903] AC 114.
5   [1966] 1 Lloyd's Rep 1.
6   *Re National Coffee Palace Co, ex p Panmure* (1883) 24 ChD 367.

In the simple case of an unauthorised contract of sale or purchase, the appropriate amount of damages will normally be the difference between the contract price and the market price,[1] assessed at the date on which completion should have taken place.[2] In addition to this basic sum, the third party is entitled to recover any foreseeable incidental expenditure. Thus the costs of taking an abortive legal action against the principal,[3] or of investigating the principal's title to land,[4] are recoverable from the agent, as are the costs of defending an action brought by the principal's solicitor without authority.[5] However, the fact that a purchaser has already agreed to resell the property concerned, and is thus liable to the sub-purchaser for breach of contract, is regarded as too remote a consequence for the agent to be liable,[6] except where the agent actually knows of the proposed resale.[7]

Termination of agency

We now turn to consider the various ways by which an agency relationship may be brought to an end, and the effect which this has on the rights and oblations of the parties.

Termination by agreement

The agency relationship is based upon agreement of the parties, and it is their agreement (properly interpreted) which primarily determines its duration. In *Danby* v *Coutts & Co*,[8] for example, the plaintiff executed a power of attorney which, after reciting that the plaintiff was going abroad for a time and wished representatives to act for him during his absence, appointed two persons to be his attorneys. It was held that the authority conferred upon these attorneys was only operative while the plaintiff was abroad; he was accordingly not bound by a fraudulent transaction carried out by the attorneys after his return.

---

1   *Simons* v *Patchett* (1857) 7 E & B 568.
2   *Chitholie* v *Nash & Co* (1973) 229 EG 786.
3   *Collen* v *Wright* (1857) 8 E & B 647.
4   *Godwin* v *Francis* (1870) LR 5 CP 295.
5   *Yonge* v *Toynbee* [1910] 1 KB 215.
6   *Spedding* v *Nevell* (1869) 4 CP 212.
7   *CH Rugg & Co Ltd* v *Street* [1962] 1 Lloyd's Rep 364.
8   (1885) 29 ChD 500.

Where there is no reason to limit an agent's authority to a certain period of time, it nevertheless comes to an end when the agent completes the task for which he or she was appointed. This means, for example, that a selling agent's authority terminates at the moment the sale is completed; the principal will accordingly not be bound by a subsequent agreement between the agent and the purchaser to alter the terms on which payment is to be made[1] or to rescind the sale.[2]

Termination by unilateral action

Agency, in common with other relationships which involve the provision of personal services, will not be enforced in court by the grant of an order of specific performance.[3] As a result, while a party who withdraws from the relationship in breach of contract may have to pay damages, the withdrawal itself will none the less be effective.[4] Indeed, the courts will not lightly impose liability at all, even in the form of damages, for there is a presumption that an agency relationship is terminable at the will of either party, even where it was originally created by deed[5] or for valuable consideration.[6]

The presumption of terminability is especially strong in the case of agents who are employed solely upon a commission basis.[7] Where, however, the agency bears a closer resemblance to a contract of employment, the agent will normally be entitled to a period of notice of dismissal (which means in effect that the principal must pay compensation based on what the agent would have earned during that period).[8] Thus, for example, in *Martin-Baker Aircraft Co Ltd v Murison,*[9] where the plaintiff manufacturers of ejector seats appointed the defendant as their 'sole selling agent for all their products ... on the North American Continent', it was held that the defendant was

---

1   *Blackburn* v *Scholes* (1810) 2 Camp 341.
2   *Nelson* v *Aldridge* (1818) 2 Stark 435.
3   *Clarke* v *Price* (1819) 2 Wils Ch 157; *Chinnock* v *Sainsbury* (1860) 30 LJ Ch 409.
4   *Whitwood Chemical Co* v *Hardman* [1891] 2 Ch 416; *Page One Records Ltd* v *Britton* [1967] 3 All ER 822.
5   *Bromley* v *Holland* (1802) 7 Ves 3, 28 *per* Lord Eldon LC.
6   *Chinnock* v *Sainsbury* (1860) 30 LJ Ch 409.
7   *Alexander* v *Davis & Co* (1885) 2 TLR 142; *Motion* v *Michaud* (1892) 8 TLR 253.
8   *Parker* v *Ibbetson* (1858) 4 CBNS 346; *Barrett* v *Gilmour & Co* (1901) 17 TLR 292.
9   [1955] 2 QB 556.

entitled to 12 months' notice of termination of this agreement. However, an agent who would otherwise be entitled to notice will forfeit that entitlement, and may thus be instantly dismissed, in cases of serious misconduct such as taking bribes,[1] and possibly even where the agent is guilty of causing loss to the principal through negligence.[2]

It should be noted that, even where an agency is terminable at will, this does not mean that the termination can take effect before the agent is notified of it. Indeed the presumption is very much to the contrary.[3] However, this presumption is subject to any contrary intention, express and implied, and it is clear that estate agents form an important exception to the general rule. It was accepted by the Court of Appeal in *EP Nelson & Co v Rolfe*[4] that the authority given to an estate agent, to find someone willing to purchase the client's property, would automatically terminate if the client exchanged contracts with another purchaser; hence, the estate agent would be unable to earn commission on an introduction made after that time, even if the agent was unaware that the property had already been sold.

Termination by operation of law

An agency relationship may be brought to an end automatically by certain supervening events, which serve to incapacitate the principal or the agent in either a legal or a practical sense. In the case of the principal, such events include death (whether known to the agent or not),[5] and winding up, which is the equivalent of death where the principal is a limited company.[6] The mental incompetence of the principal will also operate to terminate agency.[7] It should be noted that, under the Enduring Powers of Attorney Act 1985, it is now possible for a principal to execute a power of attorney which will survive the principal's subsequent insanity.[8] To achieve this effect, various prescribed formalities must be complied with and the agent,

---

1   *Boston Deep Sea Fishing & Ice Co v Ansell* (1888) 39 ChD 339.
2   See *Cussons v Skinner* (1843) 11 M & W 161.
3   *Re Oriental Bank Corporation, ex parte Guillemin* (1884) 28 ChD 634.
4   [1950] 1 KB 139. See also *AA Dickson & Co v O'Leary* [1980] 1 EGLR 25: p 100.
5   *Blades v Free* (1829) 9 B & C 167.
6   *Salton v New Beeston Cycle Co* [1900] 1 Ch 43.
7   *Yonge v Toynbee* [1910] 1 KB 215.
8   For the mental capacity required to execute such a power, see *Re K*; *Re F* [1988] Ch 310.

on realising that the principal is becoming mentally incapable, must register the power of attorney with the Court of Protection, having first given notice to certain of the principal's relatives.

Where the principal becomes bankrupt, this will normally bring the agency to an end,[1] although it will not deprive the agent of authority formally to complete a transaction already binding on the principal.[2]

As for the agent, death[3] and mental incompetence are again established as terminating events. However, the agent's bankruptcy will only be so where its effect is to prevent the agent from carrying out what he or she was appointed to do.[4]

Apart from these specific examples of personal catastrophe, a contract of agency will be terminated under the doctrine of frustration wherever its continuance would be impossible, illegal or completely impracticable. Thus, when war is declared, the agent's conscription[5] or internment as an enemy alien may have this effect, provided that it is sufficiently long-term to frustrate the purpose of the agency.[6]

Limitations on termination

Where an agent is given authority, by deed or for valuable consideration, in order to effect a security or protect an interest of the agent, the authority cannot be revoked by the principal while the security or interest continues in existence.[7] Furthermore, the authority in these circumstances will not be terminated by the death, insanity or bankruptcy of the principal.[8] However, authority is only irrevocable on this basis where it is given specifically to protect the agent's interest; it is not sufficient merely to show that the agent was owed money by the principal at the time.[9]

The common law rules described above are largely repeated in the Powers of Attorney Act 1971, section 4. However, this provision is

---

1   *Dawson* v *Sexton* (1823) 1 LJOS 185.
2   *Dixon* v *Ewart* (1817) Buck 94.
3   *Farrow* v *Wilson* (1869) LR 4 CP 744.
4   *McCall* v *Australian Meat Co Ltd* (1870) 19 WR 188.
5   *Marshall* v *Glanvill* [1917] 2 KB 87.
6   *Nordman* v *Rayner & Sturges* (1916) 33 TLR 87.
7   *Gaussen* v *Morton* (1830) 10 B & C 731.
8   *Alley* v *Hotson* (1815) 4 Camp 325; *Carter* v *White* (1883) 25 ChD 666.
9   *Raleigh* v *Atkinson* (1840) 6 M & W 670; *Smart* v *Sandars* (1848) 5 CB 895.

of somewhat narrower scope, in that it only applies to authority given in the form of a deed and only where that authority is actually expressed to be irrevocable. If these conditions are not met, there appears to be nothing to prevent the agent from relying on the principles of common law.

Effects of termination

Although the termination of an agency relationship in any of the ways described above will serve to deprive the agent of any express or implied authority to act on the principal's behalf, a third party who is unaware of the terminating event may continue to rely on the agent's apparent authority. This is certainly so where the termination results from the parties' actions[1] or from the principal's insanity,[2] although it seems that apparent authority too will cease to operate on the principal's death.[3] Further protection for innocent third parties, at least where the agent's authority is conferred by deed, is given by the Powers of Attorney Act 1971, section 5(2).

Termination may bear harshly upon an agent who, being unaware of it, continues to act. If the agent's actions are binding on the principal under the doctrine of apparent authority, then the agent will be liable for any loss caused on the basis that, as between principal and agent, the act was unauthorised.[4] Conversely, if the principal is not bound, then the agent will be personally liable to the third party for breach of warranty of authority.[5] However, statute has again provided some protection for the innocent, at least those whose authority is conferred by deed. Section 5(1) of the Powers of Attorney Act 1971 provides:

> A donee of a power of attorney who acts in pursuance of the power at a time when it has been revoked shall not, by reason of the revocation, incur any liability (either to the donor or to any other person) if at that time he did not know that the power had been revoked.

As to the rights which are normally enjoyed by agents, these are presumed to cease at the moment of termination, so that the agent

---

1   *Trueman* v *Loder* (1840) 11 A & E 589.
2   *Drew* v *Nunn* (1879) 4 QBD 661.
3   *Blades* v *Free* (1829) 9 B & C 167.
4   *Re Overweg, Haas* v *Durant* [1900] 1 Ch 209.
5   *Yonge* v *Toynbee* [1910] 1 KB 215: see p 31.

will not be entitled to remuneration,[1] or even to an indemnity,[2] in respect of transactions effected after that time. However, termination cannot deprive an agent of rights which have already accrued, such as the right to be reimbursed for expenses incurred on the principal's business.[3]

The question of accrued rights has given rise to particular difficulties in relation to those selling agents whose contracts entitle them to 'continuing commission' in respect of orders placed after the termination of their agency by customers whom they have introduced. In deciding whether such a right exists, the courts pay great attention to the actual words of the agreement[4] and appear generally reluctant to imply a term to this effect where none is expressed.[5] Indeed, even express mention of 'repeat orders' is not conclusive, since that could mean those repeat orders placed prior to the termination of the agency.[6] However, it may be noted that, where an agent is entitled to continuing commission, this can be a most valuable right, since the entitlement may last until the 'crack of doom'.[7]

Sub-agency

Authority to appoint sub-agents

The common law maxim *'delegatus non potest delegare'* indicates that, as a general rule, an agent may not entrust performance of the agency to any other person.[8] This is based on the law's view of agency as a relationship which involves some trust placed in the agent by the principal. In consequence, as was stated by Buckley J in *Allam & Co Ltd v Europa Poster Services Ltd*:[9]

> Where the principal reposes no personal confidence in the agent the maxim has no application, but where the principal does place confidence

---

1   *Farrow v Wilson* (1869) LR 4 CP 744.
2   *Pool v Pool* (1889) 58 LJP 67.
3   *Chappell v Bray* (1860) 6 H & N 145.
4   In the light of the course of dealing between the parties: *Roberts v Elwells Engineers Ltd* [1972] 2 QB 586.
5   See *Sellers v London Counties Newspapers* [1951] 1 KB 784.
6   *Crocker Horlock Ltd v B Lang & Co Ltd* [1949] 1 All ER 526.
7   *British Bank for Foreign Trade Ltd v Novinex Ltd* [1949] 1 KB 623.
8   *De Bussche v Alt* (1878) 8 ChD 286.
9   [1968] 1 All ER 826, 832.

in the agent, that in respect of which the principal does so must be done by the agent personally unless either expressly or inferentially he is authorised to employ a sub-agent or to delegate the function to another.

Whether or not an agent is entitled to delegate depends, therefore, on the scope of the agent's authority. If the principal has expressly authorised such a procedure it will obviously be valid; so too where the principal ratifies an unauthorised delegation.[1] Apart from this, authority to delegate may in certain circumstances be implied from the nature of the transaction (for example where the acts concerned are purely ministerial and involve no element of skill or discretion),[2] or from the conduct of the parties. Thus, for example, a principal who appoints an agent to carry out certain tasks, knowing that the agent does not intend to act personally, will be taken to have authorised the appointment of a sub-agent.[3] Authority may also be implied on the basis of what is usual in a particular trade or profession, although this does not appear to apply in the case of an estate agent.[4] Finally, there seems no reason to doubt that, in principle at least, delegation may be justified in an emergency,[5] although, as noted earlier, the doctrine of 'agency of necessity' is one of extremely limited scope.[6]

Legal position of sub-agents

A delegation which is unauthorised produces no legal relationship whatsoever between principal and sub-agent. Thus, where the third party seeks to take action in respect of anything done by the alleged 'sub-agent', it is the main agent and not the principal who will be held responsible.[7] Furthermore, the sub-agent in such circumstances will have no claim against the principal for remuneration,[8] and cannot exercise any kind of lien over the principal's goods.[9]

---

1   As in *Keay* v *Fenwick* (1876) 1 CPD 745.
2   Such as signing a notice to terminate an agreement: *Allam & Co Ltd* v *Europa Poster Services Ltd* [1968] 1 All ER 826.
3   *Quebec & Richmond Railway Co* v *Quinn* (1858) 12 Moo PCC 232; *De Bussche* v *Alt* (1878) 8 ChD 286.
4   See pp 74–77.
5   See *De Bussche* v *Alt* (1878) ChD 286, 310–11.
6   See p 8.
7   *Maloney* v *Hardy and Moorshead* (1970) 216 EG 1582.
8   *Schmaling* v *Tomlinson* (1815) 6 Taunt 147.
9   *Solly* v *Rathbone* (1814) 2 M & S 298.

Where an agent *is* authorised to appoint a sub-agent, such delegation may take effect in one of two ways. In the first place, the agent may be authorised to create privity of contract between principal and sub-agent, so that the person appointed becomes an agent of the principal in the full sense and will take on all the appropriate rights and duties.[1] An appointment of this kind may operate in addition to the existing agency or in substitution for it; in the latter case the original agent, having carried out the principal's instructions, drops out of the picture. It should be noted, however, that the courts are not in favour of this type of delegation, and will require clear evidence that this is what was intended. As was pointed out by Wright J in *Calico Printers' Association Ltd* v *Barclays Bank*:[2]

> To create privity it must be established not only that the principal contemplated that a sub-agent would perform part of the contract, but also that the principal authorised the agent to create privity of contract between the principal and the sub-agent, which is a very different matter requiring precise proof.

The second type of delegation, and the one which the courts will presume to have been intended, is where the sub-agent is merely an 'agent of the agent' and has no personal contractual relationship with the principal. Such a sub-agent cannot be held liable to the principal in respect of any of the normal 'agency' duties which are based upon contract, for there is no contract between them. This has led the courts to rule that the sub-agent is not directly liable to the principal for money received on the principal's behalf; the sub-agent's duty is to account to the agent,[3] and the agent is in turn liable to the principal.[4]

There seems no logical reason why a sub-agent should not incur direct liability to the principal in ways which are not based on contract (for example in the tort of negligence, or for breach of fiduciary duty).

---

1   *De Bussche* v *Alt* (1878) 8 ChD 286; *Powell & Thomas* v *Evan Jones & Co* [1905] 1 KB 11; *Foalquest Ltd* v *Roberts* [1990] 1 EGLR 50.
2   (1931) 145 LT 51, 55.
3   *Stephens* v *Badcock* (1832) 3 B & Ad 354.
4   *Mackersy* v *Ramsays, Bonars & Co* (1843) 9 C & F 818. See also *Swire* v *Francis* (1877) 3 App Cas 106; *Edwards Real Estate* v *Bamtor* (1978) 11 AR 589.

However, the case law on these issues, much of it fairly elderly, is somewhat equivocal.[1]

As far as the rights of a 'true' sub-agent are concerned, it may safely be assumed that any claims for remuneration or indemnity must lie against the agent rather than the principal. In pressing these claims, however, the sub-agent may well be able to exercise a lien over goods of the principal, although this will not apply where the appointment of the sub-agent was unauthorised.[2] The extent of an authorised sub-agent's lien depends upon whether or not the sub-agent was aware of his or her status at the time of appointment. If the existence of the principal was not disclosed, then the sub-agent may exercise any lien, whether particular or general, to the same extent as this could have been exercised against the agent.[3] If the principal's existence was disclosed, on the other hand, then the sub-agent may still exercise a particular lien;[4] any general lien, however, may not exceed the scope of the lien (if any) which the agent could have exercised against the principal.[5]

---

1   As to negligence, see *Calico Printers' Association Ltd* v *Barclays Bank* (1931) 145 LT 51; *Balsamo* v *Medici and Morris* [1984] 2 All ER 304. As to fiduciary duties, contrast *New Zealand & Australian Land Co* v *Watson* (1881) 7 QBD 374 and *Powell & Thomas* v *Evan Jones & Co* [1905] 1 KB 11.

2   *Solly* v *Rathbone* (1814) 2 M & S 298.

3   *Mann* v *Forrester* (1814) 4 Camp 60; *Westwood* v *Bell* (1815) 4 Camp 349.

4   *Fisher* v *Smith* (1878) 4 App Cas 1.

5   *Ex p Edwards* (1881) 8 QBD 262.

Duties of Estate Agents and Auctioneers

Duties to the client

Duty to act

General

The relationship which exists between estate agent and client is, for the most part, treated in law as a normal example of agency for the purpose of determining their mutual rights and duties. However, it is a more loosely-knit relationship than that of master and servant, or even that of salaried agent and principal, and this has led to suggestions that not all the normal incidents of agency are applicable. In particular, there is some disagreement as whether an estate agent who takes on a property is obliged to do anything at all to attempt to secure a sale. The view that an estate agent is completely free to act or not is exemplified by *dicta* from the decision of the House of Lords in *Luxor (Eastbourne) Ltd* v *Cooper*.[1] The best known remarks, in which are considered the basic nature of commission contracts in general and those of estate agents in particular, come from the speech of Lord Russell of Killowen:[2]

> Contracts by which owners of property, desiring to dispose of it, put it in the hands of agents on commission terms, are not (in default of specific provisions) contracts of employment in the ordinary meaning of those

---

1   [1941] AC 108.
2   At p 124.

words. No obligation is imposed on the agent to do anything. The contracts are merely promises binding on the principal to pay a sum of money upon the happening of a specified event, which involves the rendering of some service by the agent. There is no real analogy between such contracts, and contracts of employment by which one party binds himself to do certain work, and the other binds himself to pay remuneration for the doing of it.

Lord Romer said:[1]

The respondent was not employed by the appellants to find a purchaser. He was not employed to do anything at all, and would have committed no breach of his agreement with the appellants had he remained entirely inactive.

Finally, Viscount Simon LC said:[2]

I doubt whether the agent is bound, generally speaking, to exercise any standard of diligence in looking for a possible purchaser. He is commonly described as 'employed': but he is not 'employed' in the sense in which a man is employed to paint a picture or to build a house, with the liability to pay damages for delay or want of skill.

According to the view thus forcibly expressed, the true analogy with an estate agent's mandate is the offer of a reward in return for a specific act. A person is under no duty to seek the reward but, if the act is done, the reward is payable.[3] If this is correct, it leads to the conclusion that the normal relationship between estate agent and client consists of a 'unilateral contract', that is to say, one which does not ripen into a contract in the full sense until the agent 'accepts' the client's 'offer' by bringing about whatever event is specified.

Such a conclusion would have a number of unfortunate consequences. In the first place, it would mean that the terms of the 'offer' are fixed at the moment when the client first instructs the agent to act. As a result, a 'confirming letter' sent by the agent which in fact alters those terms (or the terms which are implied by law in the absence of express agreement) would take effect as a counter-offer and would thus not be binding unless specifically accepted by the client.[4]

---

1   At p 153.
2   At p 117.
3   See Murdoch: 'The Nature of Estate Agency' (1975) 91 LQR 357.
4   See p 141.

Secondly, it would cause severe difficulties of interpretation in relation to section 18 of the Estate Agents Act 1979. This requires estate agents to notify their clients in advance of all potential charges, and was clearly drafted on the assumption that the relationship is contractual from the outset.[1] Thirdly, the statutory duty of care which is imposed by section 13 of the Supply of Goods and Services Act 1982 is expressed in such terms that it can only apply once a contract for the supply of services is in existence. Therefore if the relationship consists of a 'unilateral contract', it lies outside the scope of this provision.

Since what we may call the *Luxor* v *Cooper* view has so many inconvenient implications, it is somewhat reassuring to find that it has not passed wholly without challenge. In the case itself, Lord Wright described the estate agent as having introduced a prospective purchaser 'in pursuance of his undertaking' and further stated:[2]

> It is well known that in the ordinary course a property owner intending to sell may put his property on the books of several estate agents with each of whom he makes a contract for payment of commission on a sale.

Such comments suggest that an estate agent may not be quite as free to remain inactive as the majority of their lordships appear to have thought, and a similar line was taken by two members of the Court of Appeal in the case of *Prebble & Co* v *West*.[3] In that case a vendor of property made a claim against her estate agents for negligence in relation to the sale. Counsel for the agents, relying on *dicta* from *Luxor* v *Cooper*, contended that, since an estate agent could not be sued for doing nothing at all, there must equally be immunity for doing something and doing it negligently. This proposition was described as 'startling' by Edmund Davies LJ, who asserted that if an estate agent, who had been told that the client required a quick sale, received an offer for the property at the asking price, it would be the agent's duty to pass on this offer to the client. Lord Denning MR agreed with this illustration, and said of an estate agent: 'So long as he has the house on his books, it is his duty to have regard to the interest of his client and not to do anything contrary to it.'

A further attack on the 'unilateral contract' concept of estate agency[4] uses as ammunition the earlier decision of the Court of Appeal

---

1   See p 293.
2   At p 139.
3   (1969) 211 EG 831.
4   By C R McConnell (1983) 265 EG 547.

in *Keppel* v *Wheeler*.[1] In that case a firm of agents whose commission was not earned until matters reached at least the stage of exchanging contracts were held liable to pay damages to their client for failing to disclose certain information before that date. The agents' liability could in all probability have been based upon breach of their duty of loyalty[2] (a duty which exists whether or not there is a contractual relationship between agent and client[3]) but the Court of Appeal appears to have treated them as guilty of a breach of contract; thus the decision as it stands is undeniably in conflict with the remarks of the House of Lords in *Luxor* v *Cooper*.

Sole agency

In the light of the foregoing it is impossible to say with any degree of certainty whether an estate agent may be liable for damaging a client's interests through inaction. However, it may be more confidently suggested, for the reasons outlined below, that such liability could indeed be incurred by a 'sole agent'.

It is well established that a client who appoints a 'sole agent' will be liable for selling the property through another agent. Likewise, a client who gives an agent 'sole right to sell' will be liable for selling the property privately.[4] The legal basis of the vendor's undoubted obligation in such cases has seldom been the subject of judicial investigation, but such authority as there is suggests that it rests upon some consideration, in the form of a counter-obligation, moving from the agent. In *Bentall, Horsley & Baldry* v *Vicary*,[5] for example, the agency agreement provided that, in the event of a sale, the expenses incurred in advertising the property should be borne by the client; if no sale resulted, on the other hand, this cost should fall upon client and agent in equal shares. McCardie J held that the arrangement provided 'ample consideration' to render a sole agency enforceable against the client, but it is suggested that this reasoning is suspect. Since, as we

---

1   [1927] 1 KB 577.
2   As it was in the Canadian case of *Jackson* v *Packham Real Estate Ltd* (1980) 109 DLR (3d) 277.
3   See p 62.
4   *Chamberlain & Willows* v *Rose* (1924) summarised at [1931] 1 KB 261; *Hampton & Sons Ltd* v *George* [1939] 3 All ER 627: p 126.
5   [1931] 1 KB 253.

shall see,[1] the client of an estate agent is not usually responsible for the agent's expenses unless this is specifically agreed, the special arrangement in *Bentall*'s case was in truth a concession by the client and could hardly therefore be said to constitute consideration on the agent's part.

In cases where there is no special arrangement of the kind found in *Bentall*'s case, judges have on occasion found the consideration which they regard as necessary to support a sole agency in an implied undertaking by the agent to use 'best endeavours' to bring about a sale. The two cases most commonly cited in this connection are *E Christopher & Co v Essig*[2] and *Mendoza & Co v Bell*.[3] Unfortunately, neither of these decisions is fully reported and, moreover, certain remarks attributed to Goddard LJ in the latter case cast some doubt upon his line of reasoning. In reply to his own question: 'Why should a property owner appoint a sole agent?', his lordship is said to have explained:

> If he employed a sole agent it seemed ... that the contract must be that the agent would do his best to find a purchaser and would be committing a breach of his contract if he did not do something, although he might not be successful. Otherwise a property owner was getting no benefit by appointing a sole agent.

Since the vendor in this case had claimed that it was precisely because he was getting no benefit from the appointment of a sole agent that he should not be liable for the breach of the sole agency agreement, it may be objected that the judge's reasoning assumes the very thing which it sets out to prove.

The idea of an implied obligation on the part of a sole agent to use 'best endeavours' is open to two further objections. First, while such an obligation has been examined by the courts in situations beyond the world of property,[4] it is difficult to see how it would operate in the field of estate agency. No doubt some form of positive action would be expected of the agent, but what precisely would this involve? In *Glentree Estates Ltd v Gee*,[5] where the client of a sole agent suggested

---

1   P 150.
2   [1948] WN 461.
3   (1952) 159 EG 372.
4   *B Davis Ltd v Tooth & Co Ltd* [1937] 4 All ER 118; *Ault & Wiborg Paints v Sure Service* [1983] *The Times*, July 2.
5   [1981] 2 EGLR 28.

that the agent was under a positive duty to advise as to price and sale, to provide regular 'progress reports' and to negotiate with applicants in an effort to raise the price, Ewbank J expressed strong doubts as to whether any such duties were owed. Secondly, it would be extremely difficult for a client to prove any loss flowing from an agent's failure to use best endeavours. Unlike a deprived sole agent, who can always point to the fact that the property has actually been sold as showing that the agent too could have sold it, a frustrated vendor has no such convenient evidence available.

The uncertainty of this area of the law has led many agents to make some express promise in an effort to ensure that their sole agency is enforceable. Such a promise may be in general terms, eg to use one's best endeavours, or it may consist in the provision of some special service denied to ordinary clients. In either case, it appears that an express promise takes the place of any obligation which would otherwise be implied. This, at any rate, is a conclusion which may be drawn from the decision of the Court of Appeal in *Midland Business Agency* v *Apted*,[1] where the claimants were granted the sole right to sell the defendant's business in return for their inclusion of the business in their 'sales circulation scheme'. Upon proof that there was no such special scheme, it was held that consideration for the sole agency had totally failed and it was, accordingly, unenforceable by the claimants. Although the point was not expressly taken, the decision appears to bear the implication that, if an express undertaking is not honoured, the agent cannot seek to support sole agency rights by reverting to an implied promise, such as that the agent will use best endeavours on the client's behalf.

In spite of the difficulties inherent in identifying any positive obligation owed by a sole agent to the client, let alone in discovering precisely what, in practical terms, such an obligation entails, there is no doubt that the English courts hitherto have treated the legal basis of sole agency as being a bilateral contract. It is therefore of interest to note that a Canadian judge has expressed the firm view that sole agency and multiple agency alike consist of unilateral contracts.[2] Since a client who commits a breach of a sole agency agreement is liable under Canadian law just as in England,[3] this view may be met with the objection that

---

1   (1971) 218 EG 1727.
2   *Bradley-Wilson (1954) Ltd* v *Canyon Gardens Ltd* (1965) 53 WWR 413, 417, *per* Sheppard, J A.
3   *Galan* v *Alenko* [1950] 3 DLR 9; *Fidelity Trust Co* v *Rudy* (1957) 23 WWR 668.

there can be no liability for the mere withdrawal of an offer at a time when that offer has not yet been accepted. It is submitted, however, that, since the English law which governs unilateral contracts already recognises that an implied promise by the offeror not to revoke an offer becomes legally binding once the offeree acts upon it, there is no reason whatsoever why the client's express promise not to sell the property by other means should not become binding in exactly the same way. Such a view, if generally accepted, would ensure the continued enforceability of sole agency agreements, while freeing the law from the artificial and often fictitious constraints of 'best endeavours'.

Duty to comply with instructions

It is an agent's fundamental obligation to comply with all the client's express instructions. Thus, an estate agent or auctioneer who disregards a specific prohibition imposed by the client, or who fails to honour an explicit undertaking given to the client,[1] will be liable for any loss which is thereby caused. The only instructions which may safely be ignored are those which are unlawful, so that to carry them out would make the agent a party to the illegality. In *Bexwell* v *Christie*,[2] for example, where conditions of sale stated that 'goods should be sold to the best bidder', it was held that the auctioneer was not liable to the vendor for failing to comply with the latter's order to impose a secret reserve on a particular lot. Similarly, in *Narramore* v *Fuller, Hall & Foulsham*,[3] where reserves had validly been placed upon certain lots, the vendor appeared in the sale room and bid in person. The auctioneer, who had been instructed to bid on the vendor's behalf, thereupon refrained from doing so and, as a result, certain items were sold for less than their reserve. The auctioneer was held not liable to the vendor for, once the latter had made a bid, the auctioneer could not lawfully have joined in the bidding.

One way in which liability of this kind may arise is where an agent, despite exceeding whatever actual authority has been given, succeeds in rendering the client legally liable to a third party. For example, a client who loses a sale, or who has to pay damages, because

---

1   See *Hunt* v *Beasley Drake* [1995] NPC 35: p 55.
2   (1776) 1 Cowp 395.
3   (1932) 76 Sol Jo 289.

of an estate agent's misrepresentations, may in turn recoup all the losses from the offending agent.[1] Less commonly, an estate agent may be able to bind a client to a contract of sale, perhaps by virtue of the doctrine of apparent authority; once again, assuming that the agent had no actual authority to conclude the deal, he or she will be liable to the client for breach of duty.[2] However, everything turns on the meaning of the instructions given, so that an agent who does no more than write a letter which the client has actually authorised will not be liable to the client if this turns out to have more sweeping legal consequences than either party envisaged.[3]

Cases of the kind mentioned above are relatively rare, perhaps because an estate agent is so seldom authorised to bind the client in law. However, there are many other ways in which an estate agent's straightforward breach of duty may cause loss to the client. In *Papé* v *Westacott*,[4] for example, an agent was instructed to give a client's tenant written permission to assign his lease, provided that the tenant first paid off all arrears of rent. The agent in fact gave the licence to the tenant in return for a cheque; when this was subsequently dishonoured, the agent was held liable to account to the client for the money lost. In *Edmunds* v *Andrews & Ashwell*,[5] agents acting for the vendor of a café handed over the keys to prospective purchasers so that they could inspect the premises and stock; once in possession, these applicants refused to leave and proved somewhat difficult to remove. On the evidence before the court, the clients failed to show that they had specifically instructed the agents to retain possession of the keys; however, the judge made it quite clear that, if such an instruction had been given, the agents would have been liable for failing to obey it. And in *Benham & Reeves* v *Christensen*,[6] estate agents were told by their client that no 'For Sale' board was to be erected outside the property. The agents complied with the letter of this instruction but, having introduced an applicant with whom contracts were duly exchanged, they put up a 'Sold by' board. It was held that, while an agent would normally be entitled by custom to erect a board,

---

1   *Whiteman* v *Weston* [1900] *The Times*, March 15; *Atkins* v *A D Kennedy & Co Ltd* [1921] NZLR 977.

2   *Hayes* v *Douglas* [1976] 5 WWR 308.

3   *Graylaw Investments Ltd* v *J R Bridgford & Sons* [1983] 1 EGLR 32.

4   [1894] 1 QB 272.

5   [1982] 1 EGLR 36: p 54.

6   [1979] CLY 31.

this is subject to any express instructions from the client, so that the agents in this case were guilty of a breach of contract. Since, however, the client here had suffered nothing more than annoyance, he was entitled only to nominal damages of £5.

An auctioneer, like an estate agent, must obey without question all lawful instructions from the client. Provided that the instructions are clear, the auctioneer is allowed no discretion in deciding whether or not they are in the client's best interests. Thus, an auctioneer who was instructed to sell for ready money only was held personally liable to the vendor for accepting a bill of exchange.[1] So too, when a chattel auctioneer sold a number of items under his own descriptions, rather than the ones provided by the client, he was held liable for the difference between what the goods realised and what they should have realised if properly described.[2]

Where no express instructions have been given with regard to a particular matter, the auctioneer's duty is to act in accordance with trade usage and not to exceed whatever authority arises by implication. An act which is not expressly authorised, and which is not regarded as normal practice, may involve the auctioneer in personal liability to the client. Thus, in *Brown* v *Staton*,[3] an auctioneer who allowed a purchaser of goods to set off a debt alleged to be due from the vendor was held liable to account to the client for the full price. Similarly, in *Nelson* v *Aldridge*,[4] an auctioneer was held liable for allowing the purchaser of a horse to rescind the contract. This, said Best J, was a clear deviation from the normal course of an auctioneer's duty; it was therefore incumbent upon the auctioneer to show some justification for his actions which, in this case, he had failed to do.

Where an auctioneer is guilty of disobeying specific instructions from a client, one would expect the extent of the auctioneer's liability to be measured by the amount of the loss (if any) which the client has suffered. However, the courts have on occasion treated auctioneers rather more harshly than this in assessing damages against them. In the Canadian case of *City of Halifax* v *Miller & Johnson Auctioneers Ltd*[5] the

---

1   *Earl of Ferrers* v *Robins* (1835) 2 Cr M & R 152.
2   *Brown* v *Draper & Co* (1975) 233 EG 929. In order to avoid such liability, auctioneers' terms of business frequently give them discretion to sell under their own descriptions.
3   (1816) 2 Chit 353.
4   (1818) 2 Stark 435.
5   (1980) 40 NSR 35. The decision, it is submitted, is open to serious question.

defendants, who were instructed to sell by auction a number of vehicles belonging to the claimants, sold several of these at less than the reserve prices which the claimants had stipulated. The trial judge dismissed the claimants' claim for damages on the ground that, since they had brought no evidence to suggest that the vehicles were worth any more than they had been sold for, they had failed to show that they would have been any better off if their instructions had been obeyed. The Nova Scotia Court of Appeal, however, held (for reasons which are not entirely clear) that the claimants were entitled to recover the difference between the reserve prices and the amounts actually realised.[1]

Estate agent's duty of care

Whether or not the mere appointment of an estate agent imposes any positive obligation to look for a purchaser, there is no doubt that, once the agent takes the first step, a duty of care and skill will be owed to the client. Whether or not this duty arises under the provisions of the Supply of Goods and Services Act 1982,[2] it is well established at common law. Moreover, the agent will also owe the client a duty of care in the tort of negligence. This tortious duty will, in most cases, be coextensive with the contractual one, although it is possible for the agent to assume a responsibility which goes beyond his or her contractual obligations.[3]

Acting for vendor

In the normal case of an estate agent marketing property on behalf of the owner, the most common allegation of negligence is that the agent has led the client into a disadvantageous transaction.[4] Such allegations

---

1   See also *Hayes* v *Douglas* [1976] 5 WWR 308, in which a similar measure was adopted in the case of an estate agent who sold property at less than the price authorised by the client.

2   See p 44.

3   See the views expressed by the Court of Appeal in *Holt* v *Payne Skillington* (1995) 77 BLR 51: p 55.

4   See, for example, the unsuccessful claims in *Lough Eske Holdings* v *Knight Frank & Rutley* [1991] EGCS 18; *Muldoon* v *Mays of Lilliput Ltd* [1993] 1 EGLR 43; *Berkowitz* v *MW (St John's Wood) Ltd* [1995] 1 EGLR 29; *Costa* v *Gentra Ltd* [1996] EGCS 198; *Routestone Ltd* v *Minories Finance Ltd* [1997] 1 EGLR 123.

frequently surface after the property is sold, when they are used as a defence to the agent's claim for commission.[1] They may call into question the marketing techniques adopted by the agent,[2] or the agent's view as to the value of the property.[3] An example of the latter kind is *Watts* v *Savills*,[4] where the agent formed the opinion (wrongly, as it turned out) that the land being sold had no development value. In reversing the decision of the trial judge and ruling in favour of the defendants, the Court of Appeal emphasised that the question was not whether the agents had been right or wrong, but whether their opinion was one which no competent agent could have held.

In *John D Wood & Co (Residential & Agricultural) Ltd* v *Knatchbull*,[5] the vendor of a mews house in London claimed that his estate agents had been negligent in two respects: advising him to set an asking price of £1.5 million (at which the property was sold within two months) and failing to inform him when they learned that the asking price for another house in the same mews was £1.95 million (this sold soon after for £1.8 million). It was held that the agents were not negligent in respect of advising on the asking price, since this fell within the range that a competent estate agent might have regarded as appropriate. However, they were negligent in failing to pass on information about a significant event that might influence their client's instructions and actions. They were accordingly liable for depriving their client of the opportunity to sell his house at a higher price within the next few months, damages being assessed at an appropriate percentage of the additional sum which he might be expected to have received.[6]

The result of negligence by a vendor's estate agent will, in the vast majority of cases, be that the property is sold for less than it is truly

---

1 Such claims (all unsuccessful) arose in *Knight Frank & Rutley* v *Randolph* [1991] 1 EGLR 46; *Watson* v *Lane Fox & Partners Ltd* [1991] 2 EGLR 21 and *Letgain Ltd* v *Super Cement Ltd* [1994] 1 EGLR 43.

2 As in *Cuckmere Brick Co Ltd* v *Mutual Finance Ltd* [1971] Ch 949.

3 In *Candle Services Ltd* v *Warren Reid Meadowcroft* [1995] EGCS 8, it was held that the defendant estate agents were not guilt of negligence for failing to predict a sudden sharp fall in the commercial property market.

4 [1998] EGCS 99.

5 [2003]08 EG 131.

6 From this sum there was deducted the interest which the client earned on the price actually received, during the period until he would have sold for more, and the additional commission which would have been payable to the agents on a sale at a higher price.

worth, and damages will reflect the difference between the sale price and the true value.[1]

The result of negligence by a vendor's estate agent will, in the vast majority of cases, be that the property is sold for less than it is truly worth.[2] Unusually, in *Kenney* v *Hall, Pain & Foster*[3] an estate agent was held liable for negligently overvaluing a client's house. Believing that the agent's valuation left a substantial safety margin, the client obtained bridging finance for the purchase of another property, only to find that his own could not be sold for anything approaching the suggested price and that his venture had led him to the verge of bankruptcy.

A negligence claim of a very different kind arose in the case of *Edmunds* v *Andrews & Ashwell*,[4] where the prospective purchasers of a café, who had been given the keys by the defendant estate agents, took possession of the property and refused to move out. The judge held that the defendants were not guilty of negligence in handing over the keys, since the risk which actually materialised would be regarded as so slight as not to require any precautions to be taken against it. However, it was made clear that to take a similar risk in respect of residential property (where intruders, once installed, are difficult and expensive to remove) would be a very different matter.

An estate agent who is guilty of negligence will be liable to the client for whatever loss or damage can be proved to have resulted. Assessment of that loss may on occasion cause problems, as it did in the case of *Dunton Properties Ltd* v *Coles, Knapp & Kennedy Ltd*.[5] The defendants there were held liable for failing to inform the claimants that the sitting tenant who was about to purchase their farm intended to resell it as soon as possible. Wynn-Parry J, at first instance, awarded the claimants £2,000 on the basis that they could, by raising the tenant's

---

1   See, for example, *Bell Hotels (1935) Ltd* v *Motion* (1952) 159 EG 496; *Weedon* v *Hindwood, Clarke & Esplin* [1975] 1 EGLR 82. Note, however, the assessment based on loss of opportunity to sell at a higher price, used in *John D Wood & Co (Residential & Agricultural) Ltd* v *Knatchbull* [2003] 08 EG 131, above.

2   See, for example, *Bell Hotels (1935) Ltd* v *Motion* (1952) 159 EG 496; *Weedon* v *Hindwood, Clarke & Esplin* [1975] 1 EGLR 82.

3   [1976] 2 EGLR 29. Believing that the agent's valuation left a substantial safety margin, the client obtained bridging finance for the purchase of another property, only to find that his own could not be sold for anything approaching the suggested price.

4   [1982] 1 EGLR 36.

5   (1959) 174 EG 723.

rent, have increased the market value of the farm by this amount. The Court of Appeal, however, held that the rent could have been raised in any case and the claimants' failure to raise it was not, therefore, a consequence of the defendants' breach of duty. In view of the difficulty in surmising what action if any the claimants would have taken if the defendants had informed them of the tenant's intention, the Court of Appeal awarded a sum of £250, admittedly as a conjecture.

Acting for purchaser or tenant

In cases where an estate agent is retained by a prospective purchaser or tenant to find a suitable property, negligence may consist of a failure to obtain sufficient information to enable the client to make the right decision. In *Carreras Ltd* v *D E & J Levy*,[1] for instance, agents who were asked to find office accommodation for their clients were held to be guilty of negligence in miscalculating and thus overstating the total floor area of the premises which they recommended. On the other hand, in *GP & P Ltd* v *Bulcraig & Davis*[2] it was held that an agent acting for a prospective tenant was under no obligation to make a personal check of the planning register to ascertain the permitted use. A telephone call to the planning office was sufficient, since the clients' solicitors would be expected to carry out more formal enquiries.[3]

Exactly how far an estate agent is expected to go, in checking information provided by the vendor or landlord, is not entirely clear. In *Hunt* v *Beasley Drake*,[4] a potential purchaser agreed to pay commission for an unsolicited introduction, on condition that the agent who made the introduction gave an assurance that the published particulars of the property in question were accurate. It was held that, having thus undertaken to check the vendor's sale particulars, the agent was liable for failing to do so. However, the judge was clearly of the opinion that, without such an undertaking, the agent would not be expected to carry out this task.

---

1   (1970) 215 EG 707. In fact the clients were able to recover only nominal damages, since they were unable to prove that the rent payable under the lease was any higher than the true market rental value of the premises.
2   [1986] 2 EGLR 148; [1988] 1 EGLR 138.
3   However, it was recognized by the Court of Appeal in *Holt* v *Payne Skillington* (1995) 77 BLR 51 that, if agents took it upon themselves to advise the client as to the planning situation, they would owe a duty of care and skill in so doing.
4   [1995] NPC 35.

By contrast, in *Computastaff Ltd* v *Ingledew Brown Bennison & Garrett*,[1] the estate agents acting on behalf of the landlords of business premises quoted too low a figure for the rateable value. The prospective tenants, who were suspicious of the figure quoted, drew this to the attention of their own agents and their solicitors, but neither of them discovered the error. It was held that the tenants were entitled to hold both their agents and their solicitors responsible for negligence, and to recover as damages the full amount of the extra rates which they had to pay during the tenancy.[2]

Acting for landlord

It has long been settled that an estate agent who is instructed to find a prospective tenant for a client's property must use reasonable diligence to ascertain that the person introduced will be a proper tenant[3] and that the client will be able to recover possession at the end of the lease.[4]

Failure in this respect may render the agent liable to compensate the client for loss of rent, the cost of legal proceedings to recover possession and also the damage which a 'rogue' tenant does to the premises.[5]

What constitutes 'reasonable diligence' is of course a question of fact; further, in accordance with the normal rule in negligence cases, it is a totally objective standard by which the agent falls to be judged. In *Brutton* v *Alfred Savill, Curtis & Henson*,[6] for example, a young employee of the defendants allowed a prospective tenant, who claimed to have forgotten his cheque book, to take possession of premises without payment of a deposit or any rent in advance. Once in possession, the tenant proved extremely difficult to remove and the landlady lost some £770 in unpaid rent and the cost of legal

---

1   (1983) 268 EG 906.
2   This was on the basis that, had they known the truth, the clients would not have taken the lease. It is none the less submitted that the measure of damage is incorrect, since the lease transaction took place at market value and the rating assessment itself was a fair and reasonable one for the premises.
3   *Heys* v *Tindall* (1861) 30 LJQB 362.
4   *Hellings* v *Parker Breslin Estates* [1995] 2 EGLR 99.
5   *Murray* v *Sturgis* [1981] 2 EGLR 22.
6   (1971) 218 EG 1417. See also *Faruk* v *Wyse* [1988] 2 EGLR 26.

proceedings. It was held that, since the normal worldly estate agent would not have been taken in by this simple confidence trick, the young employee's gullibility amounted to negligence, and the defendants were therefore liable to their client for this loss.

Although the standard required of an agent is a fairly high one, the case of *Cunningham* v *Herbert Fulford & Chorley*[1] serves as a useful reminder that liability is not strict. The agents there were absolved by a County Court judge from liability for negligence in accepting a tenant who appeared wealthy but who had no bank account. The Court of Appeal thought that the agents had been 'a bit casual' but, although it might have reached a different conclusion on the facts, did not find the evidence strong enough to say that the judge had erred in law. More generally, it was held in the case of *Bradshaw* v *Press*[2] that, while an estate agent should always obtain references from a prospective tenant, there is no obligation to check the references so obtained.

Auctioneer's duty of care

Scope of the duty

An auctioneer, like any professional person, has an obligation to manage the affairs of clients with the care and skill to be expected of a reasonably competent member of that profession.[3] As Lord Ellenborough CJ said in *Denew* v *Daverell*:[4]

> I pay an auctioneer, as I do any other professional man, for the exercise of skill on my behalf which I do not myself possess; and I have a right to the exercise of such skill as is ordinarily possessed by men of that profession or business. If from his ignorance or carelessness he leads me into mischief, he cannot ask for a recompense; although, from a misplaced confidence, I followed his advice without remonstrance or suspicion.

This skeletal statement of an auctioneer's duty of care was fleshed out in the following highly instructive passage from the judgment of May J

---

1   (1958) 171 EG 285.
2   [1983] 2 EGLR 16.
3   An obligation which is now implied by the Supply of Goods and Services Act 1982, s 13.
4   (1813) 3 Camp 451, 453. The auctioneer there forfeited the right to commission on the ground of negligence; many others have had to pay damages.

in *Fordham* v *Christie, Manson & Woods Ltd*:[1]

> In my view an auctioneer, like any other professional man, is under an obligation to his client to use reasonable care and skill in and about his calling as an auctioneer and to act in accordance with the terms of his contract with his client, the intending vendor. He must, for instance, obtain the best possible price for the property being auctioned. He must not negligently miss the nod of the head, or the wink of the eye or the raise of the hand in the auction room, indicating a higher bid. He must take reasonable care to ensure that he obtains a binding contract with the purchaser for his client, the vendor. In the ordinary auction ... the binding contract is made by the fall of the hammer. In effecting that contract the auctioneer must take reasonable care that there is no uncertainty in the mind of either the man on the rostrum or the person bidding about what the one is offering and for what the other is bidding. Nor must there be any lack of reasonable care in ascertaining the amount of the final bid for which the property is knocked down. I think also that there is a clear duty on an auctioneer selling in these circumstances to establish to his satisfaction the identity of the buyer to whom the goods have been knocked down.
>
> Where there is an auction of real property ... if the terms of the auctioneer's contract with the vendor require, which will probably be reflected in the conditions of sale at the auction, to obtain the stated, usually 10 per cent, deposit.

It is clear from the passages cited that the duty of care applies to all work carried out by an auctioneer on behalf of the vendor. In practice, however, the cases which have arisen on this part of the law fall into a number of groups, each applying to a particular area of an auctioneer's work. These groups of cases may now be considered, although it must be borne in mind that the list is in no way exhaustive.

Making a binding contract of sale

If there is one of an auctioneer's normal functions which can be regarded as fundamental, it is surely that of securing a binding and enforceable contract for the sale of the client's property. If, due to the auctioneer's carelessness or lack of skill or knowledge, the sale is ineffective, then the vendor will be entitled to seek compensation. In *Denew* v *Daverell*,[2] for example, the auctioneer's negligence lay in

---

1 (1977) 244 EG 213, 219.
2 (1813) 3 Camp 451.

failing to insert in the particulars of sale a proviso common in all sales of that type of property, with the result that the ensuing contract could not be enforced against the purchaser. In *Jones v Nanney*,[1] it was the actual conduct of the sale which gave grounds for complaint. The auctioneer there, having offered one lot for sale, then sold others before returning to the original lot, which was sold upon altered conditions. The Court of Exchequer were of the opinion that these irregularities would have been sufficient to invalidate the sale.

A modern example is provided by the case of *Friedrich v A Monnickendam Ltd*,[2] which concerned an international auction of jewellery held in Geneva. The English auctioneer conducted the auction in French, despite having only a limited knowledge of that language, and this led to a misunderstanding as to the amount of the highest bid (which came, ironically, from an Englishman). The mistake led to the sale being set aside under Swiss law, and a majority of the Court of Appeal regarded the auctioneer as having breached the duty of care and skill owed to the client.

As we shall see,[3] the English law which governs sales by auction no longer requires written evidence in order to make such sales legally enforceable, whether they concern goods or land. When such evidence was required by law, the courts were in no doubt that it was part of an auctioneer's duty to the client to ensure that there was sufficient evidence to bind the highest bidder.[4] Presumably this will continue to represent the legal position in those jurisdictions where written evidence is still a statutory requirement.

Where an auctioneer's careless misdescription of property results in the vendor either losing the sale or becoming liable to pay compensation, the auctioneer will be responsible to the client for the loss suffered. In *Parker v Farebrother*[5] the defendants, a firm of auctioneers, prepared particulars of sales of three houses owned by the claimant. Two of the houses, which were described as having three floors, had in fact only two, a misrepresentation which caused the claimant to pay compensation to the purchaser. The auctioneers were held liable to reimburse their client for this expense.

---

1   (1824) 13 Price 76.
2   (1973) 228 EG 1311.
3   P 157.
4   *Peirce v Corf* (1874) LR 9 QB 210.
5   (1853) 1 CLR 323. See also *Cole v Christie, Manson & Woods* (1910) 26 TLR 469.

Obtaining the best available price

Although an auctioneer must not positively misdescribe property which is being sold, care must be taken to show all its advantages to prospective purchasers, in order to obtain the best possible price for the client.[1] Failure to advertise the property sufficiently, or in a form designed to attract bidders, may thus amount to negligence.[2] The conduct of auctioneers in this respect was subjected to the most intense and detailed judicial scrutiny in *Alchemy (International) Ltd* v *Tattersalls Ltd*.[3] This arose out of a special bloodstock sale of high quality racehorses which was put on at Newmarket by the defendants. A yearling colt belonging to the claimants was knocked down by the defendants for some £450,000, but the highest bidder denied having made the final bid, refused to give his name and address and left the auction. The defendants took the view that to re-auction the colt immediately, at the end of a long day's sale, would be unwise; they accordingly put the animal up for sale again two days later, when it was sold for less than one-half the previous price. After a lengthy trial involving a great deal of expert evidence, it was held that the decision taken by the defendants was one which a reasonably competent auctioneer might justifiably make; as a result, the defendants were not liable to their clients for their loss on the resale.

The case of *Luxmoore-May* v *Messenger May Baverstock*[4] concerned that nightmare of provincial art auctioneers, a 'sleeper', in this instance a pair of paintings of foxhounds which were brought to the defendant firm in a dirty and poor condition and accepted by them 'for research'. The defendants, through their fine art consultant, valued the pictures at no more than £50 the pair, and an employee also took them somewhat casually to Christie's, where nothing favourable was said about them. When the defendants put the pictures up for auction they in fact reached £840; five months later they were sold again, this time by Sothebys with full attribution to George Stubbs, at a price of £88,000.

The trial judge held that the defendants' failure to spot the 'Stubbs potential' of these pictures was negligent; however, the Court of Appeal regarded the judge as having demanded far too high a standard of a

---

1   See *Brown* v *Draper & Co* [1975] 1 EGLR 15: p 51.
2   See *Cuckmere Brick Co Ltd* v *Mutual Finance Ltd* [1971] Ch 949: p 79 and *Johnson* v *Ribbins* [1997] 1 All ER 806, where an allegation of negligence failed on the facts.
3   [1985] 2 EGLR 17.
4   [1990] 1 EGLR 21.

'general practitioner' firm, and accordingly reversed the decision. Interestingly, the Court of Appeal specifically refused to express an opinion as to whether the pictures really were painted by Stubbs!

Security for the client's interests

In addition to obtaining an enforceable contract of sale at a good price, an auctioneer must protect the client's interests in other ways. In particular, the exercise of reasonable care may well require the auctioneer to demand from the highest bidder, if not payment of the full price, then at least a deposit. In *Hibbert v Bayley*[1] for example, a ship was put up for sale upon a condition that the highest bidder should immediately sign a contract and pay a deposit. The highest bidder promised that he would return in half an hour with the deposit money and would then sign the contract; however, he disappeared without trace. In an action against the auctioneers for negligence, it was held that the latitude given to this purchaser could only have been justified if they had known him personally, or if he had given a genuine London address. However, this does not mean that an auctioneer is in a sense a guarantor of the bidder's obligations; in *Hardial Singh v Hillyer & Hillyer*,[2] where the purchaser of a house simply disappeared without trace from the saleroom, an action for negligence by the vendor failed, for there was nothing which the auctioneers could have done.

It should not be thought that, as a result of the decision in *Hibbert v Bayley*, failure by an auctioneer to insist upon a deposit in cash automatically constitutes negligence. All the circumstances of the case must be considered before it can be decided whether the auctioneer has acted unreasonably. Thus, in *Farrer v Lacy, Hartland & Co*,[3] where mortgaged property was being sold by the mortgagee, the auctioneer accepted a cheque in payment of the deposit, according to custom. The cheque was subsequently dishonoured, but it was held by the Court of Appeal that the auctioneer was guilty of no default and was therefore entitled to his charges on the abortive sale.

Not only may an auctioneer be justified in accepting a cheque in payment of the deposit; there are circumstances in which it may be reasonable to waive payment of the deposit altogether. In *Cyril*

---

1   (1860) 2 F & F 48.
2   [1979] 2 EGLR 11.
3   (1885) 31 ChD 42.

Andrade Ltd v *Sotheby & Co*[1] the claimants instructed the defendants to sell a suit of armour. A condition of the sale was that a deposit of 50 per cent could be demanded, which, if not paid, would entitle the auctioneer to put up the lot again. The armour was knocked down for £5,000 to a person who had previously bid on behalf of an American millionaire. No deposit was demanded, as it was assumed that the purchaser was again bidding as agent. Unfortunately, however, the American millionaire repudiated the bidder's authority and the bidder was unable personally to pay for the armour. The vendors accordingly sought to recover their loss by taking action against the auctioneers for failing to obtain the deposit. It was held, however, that in these circumstances it was reasonable not to insist upon the payment of a deposit, and the auctioneers were accordingly not in breach of any duty owed to their clients.

The extent of an auctioneer's duty in respect of the purchase price was carefully considered in *Fordham* v *Christie, Manson & Woods Ltd*.[2] The defendants there sold a picture on behalf of the claimant to an Italian dealer who did not actually attend the sale but instead submitted a bid in advance. The picture was never delivered to the buyer for, shortly after the sale, he challenged its authenticity and refused to pay. The claimant then sued the defendants, alleging that they should have taken positive steps on his behalf to get in the price. It was held, however, by May J that, although an auctioneer undoubtedly has a right to sue the buyer for the price of goods sold,[3] there is no implied obligation to the client to exercise this right. Nor could it be said, on the facts of this case, that the defendants were negligent in not demanding a deposit; such a procedure would be unworkable, having regard to the common practice of accepting bids from overseas buyers who are neither physically present nor represented in the saleroom.

Duty of loyalty

The agency relationship is one of trust and confidence and, as a result, certain equitable duties are imposed upon every agent. These duties are all manifestations of one general principle, that the agent must not

---

1   (1931) 47 TLR 244.
2   [1977] 2 EGLR 9.
3   There is no equivalent right on a sale of land.

allow any personal interest, or the interests of others, to conflict with the obligations which are owed to the client. Where any potential conflict arises, the agent is under a strict duty to make full and frank disclosure to the client of all relevant facts, so as to give the latter an opportunity to decide whether to continue with the proposed transaction or with the employment of the agent.[1]

The conduct demanded of an agent in this connection is of a high standard. As Martin, B said in *Salomons* v *Pender*:[2]

> The principal bargains, in the employment, for the exercise of the disinterested skill, diligence and zeal of the agent, for his own exclusive benefit. It is a confidence necessarily imposed in the agent, that he will act with a sole regard to the interests of his principal.

In *Rhodes* v *Macalister*,[3] Atkin LJ, having reiterated the basic duty of an agent, continued:

> Now that is not an impossible standard of attainment. It is laid down by the law and it is in respect of a practical matter. The remedy is a very simple one and it is well within the compass of any ordinary business man. The complete remedy is disclosure, and if an agent wishes to receive any kind of remuneration from the other side and wishes to test whether it is honest or not, he has simply to disclose the matter to his own employer and rest upon the consequences of that. If his employer consents to it, then he has performed everything that is required of an upright and responsible agent.

Although these duties are sometimes described as 'fiduciary', and although they undoubtedly have their origins in ideas of trust, it is well established that a breach of them may be committed notwithstanding that the agent has acted in good faith. Thus in *Harrods Ltd* v *Lemon*,[4] the claimants, through their estate department, were acting on behalf of the defendant in connection with the sale of her house. A prospective purchaser, who was introduced by the claimants, wished to have the property surveyed, and asked the claimants'

---

1   The duty under s 21 of the Estate Agents Act 1979 is not relevant in this context, since it requires disclosure, to third parties rather than to clients: see pp 298–302.

2   (1865) 3 H&C 639, 643, quoting from *Story on Agency*, the leading textbook at that time.

3   (1923) 29 Com Cas 19, 29.

4   [1931] 2 KB 157.

building department to undertake this. The two departments were housed in different buildings and operated quite independently. Nevertheless it was held that the claimants, by acting for both sides, were in breach of their duty to the defendant.

Since fiduciary duties arise by virtue of the agency relationship, it is to be expected that they will be co-extensive in time with that relationship. This is borne out by the decision of the Court of Appeal in *Pilkington* v *Flack*,[1] which concerned the sale of the defendant's confectionery shop. The claimant, a business broker, produced a document containing his terms of business which the defendant, who was engaged in serving customers, signed without reading it. This document, which was described by the Court of Appeal as 'scandalous', provided that the claimant should have, for 26 weeks, the sole right to sell the business for £500 and that he should receive his commission even if the business was sold by someone else. The commission stipulated was more than double the usual rate and, in addition, the claimant was to be entitled to any part of the price in excess of £500. The defendant was held bound by this document since, when it was produced for her signature, the claimant was not her agent and owed her no duty of disclosure.[2]

The moment at which an estate agent's duty of disclosure begins is fairly clear; the time at which it ceases to operate has in the past given rise to some problems. In *Keppel* v *Wheeler*,[3] the claimant instructed the defendants to find a purchaser for his house, intimating that he would consider favourably offers of £6,000 or more. The defendants almost immediately introduced a Mr Essam, who made an offer of £6,150. This offer was accepted by the claimant 'subject to contract' but, before contracts were exchanged, a Mr Daniel approached the defendants and told them that he was prepared to pay £6,760 for the house. The defendants, believing that their duty to the claimant had been fulfilled, communicated this offer, not to the claimant, but to Essam, and a resale to Daniel was arranged at a price of £6,950. The matter eventually reached the ears of the claimant, who sued the agents for the difference between the two prices. The Court of

1   (1948) 152 EG 366.
2   The position would have been different if the agent had positively misrepresented the contents of the document: see *Jaques* v *Lloyd D George & Partners Ltd* [1968] 2 All ER 187; *Homebuyers Estates (Fylde) Ltd* v *Laycock* (1981) CA unreported (May 8).
3   [1927] 1 KB 577.

Appeal held that the agents were in breach of duty,[1] since the legal relationship between estate agent and client is not terminated by an agreement made 'subject to contract', but only by the execution of a binding contract of sale. Until that time the agent need not positively look for other purchasers, but must communicate to the client any better offers which are in fact received. In addition, the agent must pass on any other relevant information which, for example, shows the property to be more valuable than was supposed, or which casts doubt upon the solvency of the purchaser.

Keppel v *Wheeler* was followed in *Dunton Properties Ltd* v *Coles, Knapp & Kennedy Ltd*,[2] where a farm was being sold to the sitting tenant. Three days before contracts were exchanged, the estate agents discovered that the tenant was contemplating an immediate re-sale but, believing that they no longer owed any duty to the owners, did not inform them of this fact. The Court of Appeal held that the information was of relevance and should have been passed to the owners; the agents were accordingly liable in damages.[3]

It appears, then, that an estate agent's duty to inform the client of offers received terminates when the property is actually sold, in the sense of contracts being exchanged.[4] However, it should not be thought that the same principle necessarily applies to all the other manifestations of the agent's duty of loyalty. For example, an agent is not permitted to make use, for personal gain, of information gained in the capacity of agent, even after the agency has terminated.[5] What is more, an agent who sees an opportunity of personal gain cannot shrug off this prohibition without making a full disclosure to the principal. In *Regier* v *Campbell-Stuart*[6] the defendant, an estate agent, was asked to find a suitable house for the claimant. The defendant bought a house for £2,000 through a nominee, from whom he then purported to buy it for £4,500. Telling the claimant only that he had paid £4,500 for the house, he then sold it to her for £5,000. When the claimant discovered the truth and sued for all the profit which the agent had wrongfully obtained, the defendant claimed that, at the time of the

---

1 Damages were assessed at £600, since it was felt that the plaintiff would have accepted Daniel's original offer of £6750.
2 (1959) 174 EG 723.
3 As to the assessment of damages, see p 54.
4 The same is true of the agent's statutory duty to pass on offers: see p 296.
5 *Robb* v *Green* [1895] 2 QB 315.
6 [1939] 1 Ch 766.

sale, he had ceased to be an agent. It was held that, although an agent is entitled to deal directly with the client once the agency has been terminated, this requires the agent to give the 'fullest possible information' to the client. No such disclosure had been made in this case, and the agent was accordingly liable to account for his profit.

Conflicts of interest

The *Regier* case is an illustration of conflict of interest at its most obvious, where an agent wishes personally to sell to, or buy from, the client. The law's view of this appears clearly from the case of *Oliver* v *Court*,[1] where the defendant and his father carried on business together as surveyors and auctioneers. The father, who was asked to value the claimant's estate, did so (upon measurements taken by the defendant) but omitted to include the value of timber. The defendant acted as auctioneer of the estate at an abortive sale and, on the following day, himself agreed to purchase the property at approximately the amount of the valuation. It was held that the claimant was entitled to have this sale set aside, not only because of the under-valuation, but also because the auctioneer had acted wrongfully in purchasing the property. Richards CB said:[2] 'I am clearly of opinion, that an auctioneer, while his employment continues, cannot purchase the estate which he is engaged to sell.'[3]

One of the leading cases in this area of law is *Salomons* v *Pender*.[4] Here the claimant, who was instructed to sell land for the defendant, obtained an offer from a company in which he had a substantial shareholding and of which he became a director just before contracts were exchanged. The claimant received nothing from the company in respect of this transaction, nor was there any evidence that the land had not realised its full price; none the less, it was held that the claimant's undisclosed interest disentitled him to commission.[5]

---

1   (1820) 8 Price 127.
2   At p 160.
3   In sales of land, an obligation to disclose any intention on the auctioneer's part to acquire an interest in the property will also arise under the Estate Agents Act 1979: see p 297.
4   (1865) 3 H & C 639.
5   See also the Canadian case of *Edwards Real Estate* v *Bamtor* (1978) 11 AR 589, where a claim to commission was defeated on proof that the (undisclosed) purchaser was the estate agent's employee.

This principle was extended in the case of *Hocker* v *Waller*[1] where the claimant, who owned a leasehold flat, instructed the defendant to find either a purchaser or a sub-tenant. The defendant was unable to do so but, since he himself was looking for a flat, it was agreed that he should take a sub-lease. It was held that, in view of the changed circumstances, commission would only be payable if this had been expressly agreed and, since there was no evidence of this, the defendant's claim failed.

The prohibition on conflicts of interest means that, in normal circumstances, an agent is not permitted to act for both parties to a transaction.[2] However, it is by no means clear that this rule applies to the practice, common in certain types of auction sale, where prospective bidders notify the auctioneer in advance of the maximum amount which they are prepared to pay for a particular lot. It might well be argued that an auctioneer who knocks down property to such a bidder, without taking the bidding up to the stated maximum, would be guilty of failing to obtain the highest possible price for the vendor, and yet such judicial authority as there is on the point seems to be in favour of this practice. As Lord Mansfield said in *Bexwell* v *Christie*,[3] an auctioneer 'may fairly bid for a third person who employs him'. May J, in *Fordham* v *Christie, Manson & Woods Ltd* was even more direct:[4]

> I am quite satisfied on the evidence that this procedure is followed time and time again with auctioneers of the reputation and size of Christies or Sothebys. People come and look at the articles for sale during the period of the view, and then leave with the auctioneers authority to bid on their behalf up to a specified limit when the auction is in progress. This is a very common practice, a practice understood throughout the whole of the auctioneering and art world and about which no criticism whatever can be or is made.

Since both these statements of opinion were *obiter* (in the first case the auctioneer did not actually bid at all; in the second he did but the vendor did not complain), the legal position cannot be regarded as settled. In consequence, a prudent auctioneer who believes that such

---

1   (1924) 29 Comm Case 296.
2   See *Harrods Ltd* v *Lemon* [1931] 2 KB 157, and also the 'secret profit' cases: pp 70–71.
3   (1776) 1 Cowp 395, 397.
4   (1977) 244 EG 213, 215.

'bids' are likely to materialise should obtain express authorisation from the client to take them.

Even where an agent is not formally appointed to act on behalf of both parties to a transaction, to favour another party at the expense of the client is a clear breach of the duty of loyalty. In *Henry Smith & Son* v *Muskett*,[1] for example, the claimants were instructed by the defendant to sell a plot of land which was ripe for development. Of the various development companies interested in the site, one promised to reinstruct the claimants on the sale of the houses which were to be built, and the claimants duly tried to persuade their client to sell to this company. In the event, the claimants' advice was not taken and the property was sold to a rival applicant. This applicant, too, had been introduced by the claimants, but it was held that their conduct in placing the interests of a prospective purchaser above those of their client was sufficient to deprive them of their right to commission.

Similar reasoning underpins the case of *Heath* v *Parkinson*,[2] where the defendant, a music seller, asked the claimant to find a purchaser for leasehold premises at a price of £2,500. It was known that large tailoring firms, who had been introduced through other channels, would pay this price but the defendant, believing that the landlord would not consent to this change of use, had rejected their offers. The claimant discovered that the landlord would in fact raise no objection and, concealing both this and the identity of his applicants, introduced a small tailoring firm which purchased the lease for £2,250. It was held that the claimant was not entitled to commission, since he had knowingly caused the defendant to sell his property for £250 less than could have been obtained.[3]

Although conflicts of interest are most likely to arise between parties on opposite sides of a transaction, they are not limited to this situation. In *Eric V Stansfield* v *South East Nursing Home Services Ltd*,[4] for example, the claimant agents were retained by no fewer than three clients, all of whom were seeking suitable premises from which to operate a nursing home. When the defendants, one of these clients, duly purchased a property to which they had been introduced by the

---

1   [1978] 1 EGLR 13.
2   (1926) 42 TLR 693.
3   It is surprising, to say the least, that the client did not counter-claim for damages in respect of this loss.
4   [1986] 1 EGLR 29.

claimants, they refused to pay commission on the ground that the claimants had not disclosed this conflict. However, it was held that, while a disclosable conflict *could* arise in such circumstances (at least once more than one client 'demonstrates a real intent to buy'), the claimants here had in fact kept the defendants informed as to the actual position.

In one sense it might be said that the very nature of estate agency brings about a potential conflict of interests. After all, it is well known that estate agents act simultaneously for a number of vendors, and there must be many occasions when a prospective purchaser might well be suited by any of a number of different properties on the agent's books. Surprisingly, perhaps, it is only comparatively recently that the potential for this to bring about conflicts of interest has been specifically acknowledged by the English courts.

In *Robinson Scammell & Co* v *Ansell*[1] the claimant estate agents, fearing that their vendor client was about to withdraw from a sale which they had arranged, went directly to the purchaser and attempted to interest him in another property. The original sale did not in fact fall through, but the vendor, on discovering what the agents had done, refused to pay their agreed commission. When the agents sued, the Court of Appeal considered that the 'conflict of interest' rule must be tempered by the knowledge that estate agents have many clients; none the less, the court was satisfied that the agents were guilty of a serious breach of duty in going directly to the purchaser without first discussing matters with their client. Even so, it was held that the agents were entitled to their commission, since the evidence clearly showed that the agents had acted throughout in good faith.[2]

The problem of multiple clients arose again in *Brent Kelly* v *Cooper Associates*,[3] where a firm of estate agents were acting for the vendors of two neighbouring properties which fronted on to a Bermudan beach. The two properties were acquired (at a total cost of 4.5 million Bermudan dollars) by the same purchaser, in order to form a family compound. The vendor of one of the properties sued the estate agents for damages, on the basis that the possibility of such a joint acquisition rendered the property more valuable and that, had the agents

---

1   [1985] 2 EGLR 41.
2   In accordance with the principle established in *Keppel* v *Wheeler* [1927] 1 KB 577: p 73.
3   [1992] 3 WLR 936.

informed him of the purchaser's intentions, he would have been able to obtain a higher price. The Privy Council, however, held that an estate agent's duty to client X cannot possibly require the agent to divulge information which is confidential to client Y and that in consequence the agents were not in breach of their duty.[1]

Secret profits

One aspect of an agent's fiduciary duty which is of particular importance is the prohibition against what is termed a 'secret profit'. This includes any financial advantage whatsoever which the agent receives, over and above the agreed remuneration from the client, provided of course that it is not received with the client's knowledge and consent. An agent who receives a secret profit, such as commission from the other contracting party, is strictly liable to account to the client for this money.[2] For example, in *Andrews* v *Ramsay & Co*,[3] an estate agent who received £20 from the purchaser of property, although he was acting on behalf of the vendor, was made to hand over this sum to his client. The same principle was applied in *Price* v *Metropolitan House Investment and Agency Co Ltd*[4] where the claimant, an estate agent who was instructed to find a purchaser for some High-Street shops owned by the defendants, learned that the local authority intended to purchase a strip of the frontage for road widening purposes. The claimant passed on this information, not to the vendor, but to a prospective purchaser, who was thereby induced to make an acceptable offer for the property. The claimant then received commission on the sale of part of the property to the local authority. The Court of Appeal held that the claimant had by his conduct not only disentitled himself to commission from the defendants but also incurred liability to hand over to them his commission on the re-sale, since this constituted a secret profit.

It is not only the payment of commission from another party which can amount to a secret profit. In *Hippisley* v *Knee Bros*[5] the claimant employed the defendants to sell some pictures and agreed to

---

1   Ironically, divulging the information would presumably have been of considerable benefit to both clients.
2   The agent may also forfeit the right to commission: pp 72–74.
3   [1903] 2 KB 635.
4   (1907) 23 TLR 630.
5   [1905] 1 KB 1.

pay, in addition to commission, 'all out of pocket expenses', including the cost of printing and advertising. The defendants received trade discounts both from the printers and the newspaper proprietors, but charged the claimant at the full rate. A Divisional Court of King's Bench held the auctioneers liable to account for the amounts of these discounts to their client. Since, however, the auctioneers had honestly believed that they were entitled by custom to this money, it was held that they were not also obliged to forfeit their commission.

The client's remedies

Where an agent is guilty of a breach of the duty of loyalty, the client is entitled to claim damages in respect of any resulting loss, but this is by no means the only legal remedy available to the client. In the first place, any transaction tainted by the breach may be set aside. This was the course taken in two Canadian cases[1] where, interestingly, the undisclosed purchaser of the client's property was not the estate agent himself, but one of his employees. In each case it was held by the court that, since the duty of loyalty does not rest upon privity of contract, it applied in full to the employee, who was accordingly not entitled to enforce the contract against the client.

As noted earlier, an agent who has made a secret profit may be forced to hand this over to the client. Furthermore, if such a secret profit also fulfils the definition of a 'bribe',[2] the client may take action against the payer of the bribe (either for damages or the amount of the bribe) as well as against the agent.[3]

'A principal is entitled to have an honest agent, and it is only the honest agent who is entitled to any commission.'[4] It follows that, where an agent is guilty of a breach of the duty of loyalty, the client may refuse to pay the agent for his or her services. Moreover, it matters not in this context whether or not the client has suffered as a result of the breach: As Scrutton LJ said in *Rhodes* v *Macalister*:[5]

---

1   *Kramer* v *Cooper* [1975] 2 WWR 1; *Palinko* v *Bower* [1976] 4 WWR 118.
2   In terms of the civil (as opposed to the criminal) law, this means no more than that the third party dealing with the client has made a payment to someone whom he or she knows to be an agent, without disclosing this to the client.
3   *Industries & General Mortgage Co Ltd* v *Lewis* [1949] 2 All ER 573.
4   *Andrews* v *Ramsay & Co* [1903] 2 KB 635, 638, *per* Lord Alverstone CJ.
5   (1923) 29 Com Cas 19, 28.

> Whether it causes damage or not, when you are employed by one man for payment to negotiate with another man, to take payment from that other man without disclosing it to your employer and getting his consent is a dishonest act.

The general legal position is clearly shown by the case of *Fullwood* v *Hurley*[1] where the claimant, a hotel broker, was approached by the defendant as an interested purchaser. The claimant gave the defendant an 'order to view' a particular hotel, which provided for a payment to the claimant in the event of a sale. The defendant bought the hotel but, on discovering that the claimant was in receipt of commission from the vendor, refused to pay the agreed sum. The Court of Appeal upheld the defendant in this refusal, and Scrutton LJ said:[2]

> An agent who wants to make two contracts for double commission must do so in the clearest possible terms and with the clearest possible information to each of his principals what he is doing, otherwise he cannot sue under an alleged agreement.

The principle thus expressed seems clear, and yet the courts in subsequent cases have permitted estate agents to recover commission where, notwithstanding a potential conflict of interest, it is plain that the client has not been in any way disadvantaged. In *Meadow Schama & Co* v *C Mitchell & Co Ltd*,[3] for example, the claimants were retained by the defendants to find them business premises in the West End of London. The claimants found a suitable property but the defendants, in the hope of paying less commission, claimed to terminate the claimants' authority and asserted that the introduction had been made by another firm (who were engaged in selling the defendants' existing premises). Fearing that they were to be 'squeezed out', the claimants reached an agreement with the vendors' agent whereby they were to receive one-third of the latter's commission. When this agreement, which was never in fact implemented, came to the notice of the defendants, they claimed that it constituted a secret profit which disentitled the claimants to commission. This argument succeeded at first instance, but was unanimously rejected by the Court of Appeal.

---

1   [1928] 1 KB 498. In *Andrews* v *Ramsay & Co* [1903] 2 KB 635 and *Price* v *Metropolitan House Investment and Agency Co Ltd* (1907) 23 TLR 630 (p 70) the agents also failed in their claims for commission.

2   At p 504.

3   (1973) 228 EG 1511.

Lord Denning MR stated that, while not wishing to derogate in any way from the principle established in *Rhodes* v *Macalister*,[1] he did not regard it as applicable to the present case. The claimants had in effect completed their work so that when, faced with the dishonesty of their principals, they made a sensible and reasonable arrangement to protect their own interests, this could not be said to conflict in any way with their duty as agents.

The *Meadow Schama* case might seem a particularly strong one from the point of view of the agents, but a similar conclusion has been reached in less extreme circumstances. In *Christie Owen & Davies* v *Brown*,[2] for instance, the claimants changed their status from sub-agents of the vendor to agents of the purchaser; in *Druce Investments Ltd* v *Thaker*,[3] two firms of agents acting for vendor and purchaser agreed to divide their combined commission equally. In each of these cases the agents' conduct was judicially described as unwise and likely to provoke suspicion, but the agents were none the less successful in their claims for commission.[4]

The decision in each of the foregoing cases appears to have been based on the court's view that the conduct of the agent in question did not amount to a breach of the duty of loyalty. In other cases, however, the courts have allowed an agent to keep his or her com-mission, notwithstanding a clear breach of duty, in circumstances where that breach has occurred without any dishonest intention on the agent's part. This relaxation of the normally harsh rules governing an agent's conduct originated in the case of *Keppel* v *Wheeler*,[5] where a firm of estate agents, believing that any duty to the client ended on the introduction of a suitable prospective purchaser, failed to pass on a subsequent higher offer. The Court of Appeal, while holding the agents liable to pay damages based on the difference between the sale price and the amount which could have been obtained, nevertheless allowed the recovery of commission, since the agents' breach of duty was based, not on an improper motive, but on a genuine misconception of their legal obligations.[6]

---

1   (1923) 29 Com Cas 19.
2   (1983, unreported).
3   [1988] 2 EGLR 229.
4   As to the position of estate agents who act for more than one prospective purchaser (and the inevitability of acting for more than one vendor) see pp 69–70.
5   [1927] 1 KB 577.
6   This was followed in *Robinson Scammell & Co* v *Ansell* [1985] 2 EGLR 41: p 69.

The position of the agents was even stronger in *Harrods Ltd* v *Lemon*,[1] where the claimants inadvertently acted for both vendor and purchaser in the same transaction. On discovering this, the claimants suggested to the defendant vendor that the purchaser should obtain an independent survey. The defendant did not accept this suggestion and completed the sale at a price which was reduced because of the claimants' adverse report on the condition of the drains. The Court of Appeal held that the claimants' breach of duty did not disentitle them to commission since the defendant had, with full knowledge of the facts, elected to continue with them as her agents.

Duty not to delegate or sub-instruct

The general duty imposed upon every agent, to perform tasks personally and not to delegate them, was discussed in Chapter 1, as were the legal relationships which may arise where delegation is authorised by the principal. There is little authority upon the application of these principles to the specific case of an estate agent, which is somewhat surprising in view of the frequent use by such agents of commission-sharing agreements. Under these arrangements, an agent instructed to find a purchaser or a property offers a proportion of the commission to another agent in return for a successful introduction, and the courts have on occasion been called upon to settle disputes arising between the two agents.[2]

The extent to which a commission-sharing agreement affects the original client was considered by the Court of Appeal in *Maloney* v *Hardy & Moorshead*,[3] in which an estate agent, without the client's knowledge, invited several other agents to co-operate in finding a purchaser. The claimant, after negotiating with one of the sub-agents, agreed 'subject to contract' to purchase the property and paid a deposit. When the sub-agent absconded, taking the deposit with her, the claimant sued both the main agent and the vendor. The Court of Appeal unanimously held that the main agent was liable for the sub-

---

1   [1931] 2 KB 157.
2   See, for example, *Davis* v *George Trollope & Sons* [1943] 1 All ER 503; *Wilkinson* v *North Suburban Properties Ltd* (1959) 174 EG 213. The former decision shows a court striving to interpret a sub-agency agreement in such a way that the sub-agent cannot earn commission unless the main agent is also entitled.
3   (1970) 216 EG 1582.

agent's default, for it was clear from the evidence that he had authorised and, indeed, expected her to receive such a deposit. The vendor, however, bore no such responsibility, for she had not authorised the appointment of any sub-agents.[1] The basis of the decision was neatly summed up by Megaw LJ:

> It is plain that no express authority was given ... to employ a sub-agent. There is no pleading or finding, nor is there any warranty in the evidence for a finding, that it is well-known practice, or a custom of the trade, general or local, or that it is a necessary and recognised incident of such agency, that the agent may employ a sub-agent with authority to collect a pre-contract deposit.

Although these remarks were based upon the evidence brought in the particular case, rather than upon any general legal rule, it is felt that the courts' general disinclination to imply authority to delegate would certainly apply to estate agents. However, this is not to say that a case could not be made out for customary authority if the evidence of usage was sufficiently strong, and this indeed is what happened in the Canadian case of *Carmichael* v *Bank of Montreal*,[2] where the court gave official recognition to the North American practice of 'multiple listing'.

It is clear from the above case that the actions of a sub-agent whose appointment has not been authorised are the responsibility of the main agent alone; the client is completely unaffected by them. Indeed, according to one decision of the Court of Appeal, this legal liability is not the only misfortune which may befall an estate agent who delegates without seeking the client's authority; such action may lead the agent to forfeit commission. In *McCann & Co* v *Pow*[3] the claimants, who were appointed 'sole agents' for the sale of the defendant's flat, instructed sub-agents without his knowledge. A prospective purchaser was sent by one of the sub-agents to view the flat and the defendant, on learning that this purchaser had not come from the claimants, naturally assumed that they could negotiate privately. When the flat

---

1 Today the action would fail anyway for lack of any implied authority to accept a pre-contract deposit: see pp 242–244.

2 (1972) 25 DLR (3d) 570.

3 [1975] 1 All ER 129; followed in *Robert Bruce & Partners* v *Winyard Developments* [1987] 1 EGLR 20. *Cf* the unreported case of *Britton Poole & Burns* v *Theodosiades* (1983), where delegation was held to have been authorised by virtue of a 'confirming letter'.

was sold to this applicant, the claimants claimed commission, but their claim was unanimously rejected by the Court of Appeal.

While the justice of this decision is not in dispute (because a vendor who believes that he or she is negotiating privately may well reduce the price to take account of the amount of commission saved) it is difficult to agree with the court in basing it upon the rule prohibiting delegation. Surely, so long as the client does not incur legal liability for the acts of a sub-agent, it is of no consequence whether potential purchasers of the property are introduced directly by the appointed agent, or indirectly through another firm. In either case, the main agent appears equally to be an 'effective cause' of the subsequent sale. It is suggested that a much better justification for refusing the agents' commission claim in this case would have been that, by failing to disclose the fact of delegation to their client, and thereby weakening his negotiating position, they were in breach of their duty of loyalty.

As far as auctioneers are concerned, the question of delegation has arisen almost exclusively in connection with the auctioneer's implied authority to sign a memorandum of the sale, so as to render the contract enforceable. As we shall see,[1] such memoranda are no longer necessary on sales of either goods[2] or land,[3] and so the matter can be briefly dealt with. The courts have treated the auctioneer in this connection as agent, not only for the vendor, but also for the highest bidder, and as being impliedly authorised by both parties to sign a binding memorandum on their behalf. Where the actual signature on the vendor's behalf was that of the auctioneer's clerk, general opinion was that the vendor would not be bound,[4] although some support could be found for the contrary view.[5] However, the legal position in respect of signature for the highest bidder was quite unequivocal. The auctioneer's authority was incapable of being delegated to the clerk, and thus the purchaser would not be bound,[6] except where there was clear evidence that the purchaser had given direct authority to the clerk to sign on his or her behalf.[7]

---

1   P 157.
2   Law Reform (Enforcement of Contracts) Act 1954, s 2.
3   Law of Property (Miscellaneous Provisions) Act 1989, ss 2, 4 and Schedule 2.
4   *Coles* v *Trecothick* (1804) 9 Ves 234.
5   *Dyas* v *Stafford* (1881) 7 LR Ir 601.
6   *Peirce* v *Corf* (1874) LR 9 QB 210; *Bell* v *Balls* [1897] 1 Ch 663.
7   *Bird* v *Boulter* (1823) 4 B & Ad 443; *Sims* v *Landray* [1894] 2 Ch 318.

Although a memorandum of the sale is no longer a statutory requirement, sales by auction generate various documents such as receipts which fall to be signed by the auctioneer. In view of the uncertainly as to the extent of the clerk's authority, if any, it is suggested that a prudent auctioneer should make a point of signing personally all such documents.

Duties to non-clients

The main legal obligations of an estate agent or auctioneer are owed to the client. However, in certain limited circumstances duties may be owed to others. Of these, the most important are the duties owed by a vendor's agent to an actual or potential purchaser.

The purchaser

It is only in exceptional circumstances that an estate agent, having introduced two parties, will be a party to the contract of sale into which they enter.[1] In consequence, any legal duties owed by the estate agent directly to the purchaser must arise independently of that contract. This rules out, among other things, any liability under the Misrepresentation Act 1967, since that Act applies only to statements made by one contracting party to another.[2] This does not mean that an agent's misstatement cannot result in personal liability, but rather that it must be on a different basis: either in tort (for negligence or deceit) or through an action by the purchaser against the vendor, followed by the vendor in turn seeking an indemnity from the agent.[3]

There are none the less certain situations in which a vendor's agent undertakes personal responsibility to the purchaser.[4] In particular, an estate agent who receives a pre-contract deposit from a prospective purchaser 'as stakeholder' is personally liable to return it on demand.[5] Indeed, this will apply even where the estate agent

---

1   *Davies* v *Sweet* [1962] 1 All ER 92 is an interesting example of the agent's personal liability.
2   *Resolute Maritime Inc* v *Nippon Kaiji Kyokai* [1983] 2 All ER 1.
3   The whole question of misdescriptions by the agent is dealt with in Chapter 5.
4   For the liability of an auctioneer who refuses to sell, where the sale has been stated to be 'without reserve', see pp 179–184.
5   *Rayner* v *Paskell & Cann* (1948) 152 EG 270.

purports to accept the deposit 'as agent' for the vendor, unless the latter has actually authorised this.[1] Similar personal responsibility attaches to an auctioneer who holds a contract deposit as stakeholder.[2]

As a general principle, any agent who claims to have authority beyond what in fact exists may be liable to a third party for breach of warranty of authority. Thus, an agent who purported to grant a lease of land on behalf of a client when not authorised to do so was held liable to pay damages to the disappointed purchaser.[3] A similar fate befell auctioneers who knocked down a client's racehorse, wrongly believing that they had been authorised to sell it.[4]

An auctioneer who sells goods is personally liable to the highest bidder to see that the contract of sale is duly carried out.[5] However, it appears that there is no equivalent liability on a sale of land, except in the rare case that the contract of sale makes it appear that the auctioneer is the vendor. Only then would the auctioneer be able to enforce the contract personally, and it is submitted that the auctioneer's rights and duties should in this respect be co-extensive.

At one time, a contract for the sale of either land or goods was legally enforceable only if there was written evidence of its terms, signed by or on behalf of the party against whom an action for enforcement was brought. The courts at that time regarded the auctioneer as having implied authority from the highest bidder to sign a memorandum on the latter's behalf, so as to enable the vendor to enforce the sale.[6] However, it appears that the auctioneer is not thereby regarded as owing the duties of a normal agent to the highest bidder.[7] In *Flint* v *Woodin*,[8] it was argued that an auctioneer who sells property in which he or she has a personal interest must, as 'agent' for the highest bidder, disclose that interest. The defendant, who had successfully bid for three houses, refused to complete the contract on the ground that they were the property of the auctioneer, a fact which

---

1   As a result of *Sorrell* v *Finch* [1977] AC 728: p 243.
2   Deposits are considered in detail in Chapter 6.
3   *Godwin* v *Francis* (1870) LR 5 CP 295.
4   *Anderson* v *Croall & Sons* (1903) 6 F (Ct of Sess) 153. The position of an auctioneer who purports to sell below the reserve price is considered at pp 174–176.
5   *Benton* v *Campbell, Parker & Co Ltd* [1925] 2 KB 410.
6   *Sims* v *Landray* [1894] 2 Ch 318.
7   The balance of authority suggests that the auctioneer owes no duty to the highest bidder to sign on behalf of the vendor: see *Richards* v *Phillips* [1969] 1 Ch 39.
8   (1852) 9 Hare 618.

had not been disclosed at the time. On the facts of the case, the defendant's delay in raising this objection clearly disentitled him from relying upon it, but in any case, Turner V-C was of the opinion that the auctioneer had broken no duty:[1]

> It is quite true, that the auctioneer is, at such a sale, the agent of the purchaser, but he is not his agent for all purposes. The auctioneer, as it seems to me, may properly hold the character of owner, without any objection being taken to the sale on that ground.

Other parties

In one situation an estate agent or auctioneer, in acting for a vendor, may have to look beyond the interests of the client. This is where a mortgage lender exercises a power to repossess and sell the mortgaged property on default by the borrower in repaying the loan.

It has long been established that the lender in these circumstances owes some duty to the mortgagor to see that a proper price is obtained. In *Cuckmere Brick Co Ltd* v *Mutual Finance Ltd*,[2] the Court of Appeal decided that the mortgagee's duty was not merely to act in good faith and without reckless disregard of the mortgagor's interests, but that it involved taking reasonable care to obtain the true market value of the property. The court also held that, if the reason why a sale took place at an undervalue was that the mortgagee's advisers (in that case a firm of auctioneers) were negligent, the mortgagee would be responsible for that negligence.[3]

Although it was not necessary in *Cuckmere Brick* to consider the legal position of the auctioneers, Cross LJ was clearly of the opinion that they would, if sued, have been held personally liable to the mortgagors for their negligence. This view, which was endorsed by the High Court of Australia in *Commercial & General Acceptance Ltd* v *Nixon*,[4] would have resulted in the imposition of liability upon a firm

---

1   At p 622. The auctioneer would now be under a statutory duty to disclose any personal interest in the property (Estate Agents Act 1979, section 21), but this does not provide any direct legal remedy for the purchaser.

2   [1971] Ch 949; followed in *Johnson* v *Ribbins* [1975] 2 EGLR 78.

3   Relying on *Wolff* v *Vanderzee* (1869) 20 LT 350 and *Tomlin* v *Luce* (1888) 41 ChD 573; (1889) 43 ChD 191).

4   (1981) 152 CLR 491.

of estate agents in *Garland* v *Ralph Pay and Ransom*;[1] however, in the special circumstances of that case the mortgagor was unable to establish that a breach by the agents of their admitted duty of care had caused any loss.

The view expressed in these cases, that a mortgage lender's professional adviser owes a duty of care to the borrower when handling the sale of the repossessed property, was enthusiastically endorsed by Buckley J in *Raja* v *Austin Gray*.[2] However, the judge's ruling on this issue was overturned by the Court of Appeal,[3] on the ground that the imposition of such a duty would not be fair and reasonable. In coming to this conclusion, which was heavily influenced by the earlier decision of Judge Jack QC in *Huish* v *Ellis*,[4] the court pointed out that the mere foreseeability that loss may be suffered is not in itself sufficient to create a duty of care. And in this particular situation there were two reasons for not creating a duty: first, it might conflict with the duty owed by the agent or other adviser to the lender, and, second, it was unnecessary, since the borrower could in any event claim against the lender for any negligence by the agent.

---

1   [1984] 2 EGLR 147.
2   [2002] 43 EG 210.
3   [2003] 04 EG 151 (CS).
4   [1995] BCC 462.

Rights of Estate Agents and Auctioneers

Remuneration

The work which is done by estate agents, valuers and surveyors on behalf of their clients falls neatly into two categories. First, it may consist of services which are of value in themselves, irrespective of whether they produce any particular end result. This would include such things as surveys, valuations, expert witness work or the representation of a client before an arbitrator or a tribunal. Second, the professional services may be geared much more closely to the achievement of a desired result, irrespective of the means used to achieve it. This would apply to the sale of property, the introduction of suitable tenant or the procurement of a loan of money.

It is usual to find that the form which remuneration takes is different for the two types of work. In relation to services pure and simple, charges are normally based on the time and trouble involved. True, where the client specifically agrees to it, or where there is a legally binding custom, the agent's fees may be based upon the value of the property concerned[1] or upon some contingency, such as a reduction in the rateable value of the client's property or the obtaining of planning permission.[2] However, the courts lean heavily against such an interpretation and consequently, if a term as to remuneration

---

1   *Wilkie* v *Scottish Aviation Ltd* 1956 SC 198.
2   Provided that the rule which prohibits 'contingency fees' in litigation work is not contravened: see *Pickering* v *Sogex Services (UK) Ltd* [1982] 1 EGLR 42; *Picton Jones & Co* v *Arcadia Developments Ltd* [1989] 1 EGLR 43.

has to be implied,[1] it will invariably be for payment of a sum to be assessed on a 'time and trouble' basis.[2]

Where what is in issue is commission, payable on the occurrence of a certain event, rather different considerations apply. In the absence of a clear express provision to the contrary, agents who work on a commission basis are not entitled to any remuneration in respect of abortive work.[3] This principle, which applies to an auctioneer as well as to an estate agent,[4] means that an agent who intends to charge a client for a valuation, if no purchaser is found for the property, must ensure that this is expressly stated in the terms of agency.[5]

Since an unsuccessful agent normally receives neither commission nor expenses, it may be said that in a sense the agent shares in the risks of the client's venture; as a result, if that venture proves successful, the agent will expect to be paid on a participation basis. Of course, this rate of payment will usually be higher than one which is geared to 'time and trouble', but this is logically inevitable. After all, if a professional adviser's time is 'worth' a certain amount, and one-half of that time is spent on work which is unsuccessful and which therefore earns nothing, it follows that the remainder must be charged at double rates. Thus, where an adviser is paid commission under a 'no sale, no pay' arrangement, the satisfied clients effectively pay for the work which is done on behalf of the unsatisfied ones.[6]

The claims of estate agents to be paid commission for their services have proved a remarkably fertile field of litigation, both at first instance and in the Court of Appeal. Many of the cases have turned on the correct interpretation to be given to the relevant agent's instructions in order to ascertain whether the event has occurred on which commission is made payable. We shall accordingly look first at this particular problem, before considering a further requirement imposed by law, namely, that the agent must be an 'effective cause' of the transaction which takes place. The special rules applicable to 'sole

---

1   For example because there is no express agreement (Supply of Goods and Services Act 1982, s 15).

2   *Upsdell* v *Stewart* (1793) Peake 255; *Drew* v *Josolyne* (1888) 4 TLR 717; *Faraday* v *Tamworth Union* (1917) 86 LJ Ch 436. See generally, Murdoch: 'Professional Fees — How Much is Reasonable?' [1981] Conv 424, 425–428.

3   See 'Remuneration where no sale results': p 144.

4   *John Meacock & Co* v *Abrahams* [1956] 3 All ER 660.

5   See *Gross Fine & Krieger Chalfen* v *Clifton* (1974) 232 EG 837.

6   As to whether this view of estate agency is shared by the courts, see pp 134–138.

agents' will be discussed, and attention will be given to the amount of commission which will be awarded in cases where this is not specified at the outset, and to the formalities with which an agent may have to comply in order to ensure that commission is legally recoverable. Finally, we shall consider certain other rights which estate agents and auctioneers may exercise against their clients.

The commission-earning event

It must be emphasised at the outset that the question whether or not an estate agent has fulfilled the client's instructions is one which can only be answered by examining and interpreting those instructions. Further, this interpretation must take place in the context of the particular case, a factor which serves to reduce the importance of previously decided cases when a claim arises for adjudication. The individuality of each case was stressed by Lord Wright in *Luxor (Eastbourne) Ltd* v *Cooper*,[1] the leading case on this subject:

> I deprecate in general the attempt to enunciate decisions on the construction of agreements as if they embodied rules of law. To some extent decisions on one contract may help by way of analogy and illustration in the decision of another contract, but, however similar the contracts may appear, the decision as to each must depend on the consideration of the language of the particular contract, read in the light of the material circumstances of the parties in view of which the contract is made.

Nevertheless, in spite of the clear warning, decisions on the construction of agreements are undoubtedly often treated, especially by litigants, as expositions of some universally valid principle, a fact of life remarked upon rather wryly by Atkin LJ in *James* v *Smith*:[2]

> As a matter of fact, the result depends so often on the terms of the contract, and nothing but the terms of the contract, that it is very seldom that one case really is a guide in respect of future cases; nevertheless, it is our experience in every commission case, and necessarily so, to find about half-a-dozen cases cited on one side, and very often half-a-dozen other cases upon the other.

---

1   [1941] AC 108, 130.
2   (1921), [1931] 2 KB 317n.

Indeed, this tendency to rely upon past cases is not without its exponents among the judiciary. As Denning LJ put it:[1]

> So many cases have now come before the courts on claims by house agents to commission that the document cannot, I think, be interpreted *in vacuo*. It must be interpreted in the light of the general law on the subject.

Before plunging into the welter of cases which comprise 'the general law on the subject', a point should be made which, if borne in mind, may help to explain, though not to justify, some of the anomalies and conflicts which will be encountered. Much of the confusion in this part of the law arises from a failure, by and large, on the part of the courts to distinguish between an agent's right to commission and the right to damages where the client wrongfully prevents the agent from earning commission. The former, by definition, depends upon the fulfilment of the agent's instructions; if these are not carried out then, however reprehensible the client's conduct, and whatever liability for breach of contract may ensue, the client cannot possibly be liable to pay commission as such.

Lest it should be thought that this is mere semantics and that it matters not what the money is called, so long as it is awarded, brief mention may be made of a possible consequence of some practical importance. It by no means necessarily follows that an award of damages for wrongfully preventing an agent from earning commission will be equal in amount to the commission itself. The damages are designed to compensate the agent for loss of an opportunity and, where there is no certainty that the opportunity would have been taken, the sum to be awarded should be reduced accordingly. However, while this method of assessment has on occasion been adopted in 'sole agency' cases,[2] the judicial tendency, in cases where an estate agent successfully sues a client, is still to speak of 'commission', thereby implying that the agent has carried out the client's instructions. As we shall see, this implication quite frequently imposes a severe strain upon the English language, and it is submitted that a simpler approach would be to extend the analysis used by Lord Russell of Killowen in *Luxor*'s case. This would involve an acknowledgment that, while 'commission' depends upon completion, damages are available to the agent whenever the client wrongfully withdraws from the transaction in question.

---

1   *Dennis Reed Ltd* v *Goody* [1950] 2 KB 277, 284.
2   See p 127–128.

Be this as it may, the purpose of this chapter is not to seek academic analyses, nor to attempt a theoretical reconciliation between judicial ideas which are patently irreconcilable, but to investigate the practical effects of various types of commission agreement. Thus, for the most part, the question: 'damages or commission?' will be, if not ignored, at least subordinated to the question: 'Is the agent entitled to be paid?'

Completion of the sale

With the foregoing in mind, we may now move to a consideration of the various forms of words used by estate agents in attempting to secure their commission. Any inquiry into this area of law must really begin with the seminal decision in *Luxor (Eastbourne) Ltd* v *Cooper*[1] of which mention has already been made and of which the facts were as follows:

The appellant companies, who wished to sell four cinemas, told the respondent that, if he introduced a person who bought the cinemas for at least £185,000, they would pay him a 'procuration fee' of £10,000 on completion of the sale. The respondent introduced a potential purchaser who made an offer of £185,000 'subject to contract' and who remained at all material times both willing and able to proceed with the transaction. The appellants, however, for reasons which are not important in this context, decided not to sell to the person introduced and so no binding contract was ever signed. It was clear from these facts that the respondent could not claim to be paid commission, for this was expressly made payable on completion, and there had been no completion. Nevertheless, the respondent argued that he was entitled to damages for breach of a term which, he said, was to be implied into the commission agreement. This term was said to be that the appellants would 'do nothing to prevent the satisfactory completion of the transaction so as to deprive the respondent of the agreed commission'.

The respondent's claim, which was accepted by the Court of Appeal, was that the appellants had broken this implied term by refusing to make and complete the contract with the willing and able person whom he had introduced. The House of Lords, however, unanimously rejected this argument and the previous decisions on which it was based.[2] The House of Lords pointed out that terms are

---

1   [1941] AC 108.
2   The Court of Appeal decisions in *Trollope (George) & Sons* v *Martyn Bros* [1934] 2 KB 436 and *Trollope (George & Sons)* v *Caplan* [1936] 2 KB 382 were overruled; *Prickett* v *Badger* (1856) 1 CBNS 296 was treated as resting on its own facts.

only implied into contracts where it is necessary to do so in order to give business efficacy to the contract. In this case no such need arose, since the contract as it stood bore the following, perfectly reasonable, construction: 'The agent takes the risk [of the sale failing to materialise] in the hope of a substantial remuneration for comparatively small exertion ... A sum of £10,000, the equivalent of a remuneration of a year's work by a Lord Chancellor, for work done within a period of 8 or 9 days is no mean reward, and is one well worth a risk.'[1] Lord Romer further criticised the cases which supported the implication of such a term on the grounds that, if the vendor was thus prevented from withdrawing from negotiations for any reason whatsoever, the implied term was wholly unreasonable; if it only prohibited the vendor from withdrawing 'without just cause or reasonable excuse' it was so vague as to be void for uncertainty.

Taken at face value, the decision in *Luxor*'s case appears to mean that an estate agent who is to be paid commission on completion will have no right of action if the sale falls through for any reason. In fact, however, a qualification referred to in two of the speeches goes a long way towards mitigating this apparent harshness. Lord Russell of Killowen, having explained his refusal to imply into the contract the term which the respondent sought, went on to say:[2]

> The position will not doubt be different if the matter has proceeded to the stage of a binding contract having been made between the principal and the agent's client. In that case, it can be said with truth that a 'purchaser' has been introduced by the agent. In other words, the event has happened upon the occurrence of which a right to the promised commission has become vested in the agent. From that moment, no act or omission by the principal can deprive the agent of that vested right.

Similarly, Lord Wright said:[3]

> If the negotiations between the vendor and the purchaser have been duly concluded and a binding executory agreement has been achieved, different considerations may arise. The vendor is then no longer free to dispose of his property. Though the sale is not completed, the property in equity has passed from him to the purchaser. If he refused to complete, he would be guilty of a breach of agreement *vis-à-vis* the purchaser. I think, as at present advised, that it ought then to be held that he is also in

---

1   *Per* Lord Russell of Killowen at p 125.
2   At p 126.
3   At p 142.

breach of his contract with the commission agent — that is, of some term which can properly be implied.

These remarks, which were made the basis of the Court of Appeal's decision in *Alpha Trading Ltd* v *Dunnshaw-Patten Ltd*,[1] mean that, in the type of commission agreement exemplified by *Luxor*'s case, the vendor will not be able to avoid paying the agent remuneration by resiling from a binding contract which the agent has procured.

Luxor's case is an example of an agreement which provides expressly for the payment of commission only on completion of the sale. The same result is reached by implication wherever reference is made to the payment of commission out of the purchase price. In *Beningfield* v *Kynaston*,[2] the claimant was instructed to find a purchaser for the defendant's estate in return for 'the usual commission out of the purchase money'. A person introduced by the claimant signed a contract and paid a deposit, only to forfeit this on his subsequent failure to complete. It was held that the true meaning of the agreement was that the agent should be paid out of the total purchase price only when this was received by the vendor.

A similar decision was reached in the case of *Beale* v *Bond*,[3] where the owner of some houses told an estate agent that he could retain as commission anything in excess of £1,150 for which the property might be sold. As in the previous case, a contract was signed which the purchaser then failed to complete, forfeiting his deposit. It was held, once again, that commission could not be paid until completion.

As a timely reminder that every case depends on its own facts and particularly upon the actual wording of the agreement, mention may be made of the case of *Passingham* v *King*.[4] Here an agent was instructed to find a purchaser of a public house in return for £100 'upon the purchase money'. The facts were very similar to those of the two cases already discussed but in this case the Court of Appeal awarded the agent his commission. A majority of the court[5] based its decision upon the ground that, in the circumstances of this contract, the words 'upon the purchase money' meant 'upon the money contracted to be paid'. Thus completion of the sale was not a pre-requisite of payment to the agent.

---

1   [1981] QB 290.
2   (1887) 3 TLR 279.
3   (1901) 17 TLR 280.
4   (1898) 14 TLR 392.
5   Vaughan Williams LJ concurred in the result on different grounds.

Find a purchaser

We may now consider a second group of commission cases in which, although the legal analysis is somewhat different from those already discussed, the practical results appear indistinguishable. These cases turn upon the construction of the phrase 'find a purchaser' and other similar instructions. In *Raymond* v *Wootton*[1] the claimant, whose sole agency for the sale of the defendant's house had expired, nevertheless kept the property on his books and introduced a person interested in purchasing it. An agreement was reached 'subject to contract' and a deposit paid. The defendant then received a higher offer from a person introduced by other agents and sold the house to him. It was held that, in the absence of a binding contract between the defendant and the person introduced by the claimant, the claimant was not entitled to commission; nor had the defendant committed any wrongful act which would render him liable in damages. A similar decision was reached in *Jones* v *Lowe*[2] where the agent was to receive commission in the event of his 'introducing a purchaser'. Hilbery J, applying the principle of *Luxor*'s case, held that this condition would not be satisfied until someone introduced by the agent 'goes so far as to sign a legal contract binding him to go through with the purchase'.

To be a 'purchaser', then, a person must sign a binding contract of purchase; however, the mere fact that this has been done is not sufficient to entitle the agent to commission under the clauses now in question. In *Poole* v *Clarke & Co*[3] the defendants, a firm of business transfer agents, were retained by the claimant in the following terms: 'I hereby instruct you to find a purchaser for my business ... I also instruct you to receive any deposit paid by such purchaser and to apply the same, on the signing of the contract, as far as possible, towards payment of your expenses and commission.' The defendants introduced someone who signed a contract to purchase the business, but who failed to raise sufficient money to complete and so forfeited the deposit which he had paid to the agents. The agents claimed to retain this deposit in part payment of their commission, but the claimant was held entitled to recover the money from them. Although a contract had undoubtedly been procured, they could not claim to have found a 'purchaser' since, as Singleton J said: 'I think it must be

---

1   (1931) 43 TLR 606.
2   [1945] KB 73.
3   [1945] 2 All ER 445.

taken that in the ordinary way the meaning of 'purchaser' in such a document is a purchaser who is willing and able to complete.'

The legal position where this particular form of words is used also applies to a number of other fairly common clauses including: 'on a sale being effected'[1] 'in the event of business resulting';[2] 'find someone to buy';[3] and 'if the property is sold'.[4] It is perhaps ironic that in *James v Smith*,[5] a Court of Appeal decision often relied upon to support the above views, both Bankes and Atkin LJJ appeared to suggest that an agent who was to receive commission 'if the business is sold' could satisfy the condition by introducing a person willing and able to purchase, even if the vendor then refused to enter into a contract with that person. Since, however, the person introduced by the agent in that case was not financially 'able' to purchase, the question did not fall to be decided; these observations were, therefore, strictly *obiter* and, in the light of subsequent decisions, may safely be ignored.

The basic rule

It may be helpful at this point to summarise the legal requirements imposed upon an estate agent by the commission agreements so far discussed. These requirements, it is submitted, are the same, regardless of whether the agreement is of the 'find a purchaser' type or the 'completion' type.

- The estate agent must introduce a person who actually enters into a legally binding contract.[6] This rule is inflexible; if the vendor declines, however arbitrarily, to contract with the person introduced, the agent can have no claim to commission or damages.

- The 'purchaser' must be both willing and able to carry on and complete the purchase.

---

1   *Martin* v *Perry & Daw* [1931] 2 KB 310.
2   *Murdoch Lownie Ltd* v *Newman* [1949] 2 All ER 783.
3   *McCallum* v *Hicks* [1950] 2 KB 271.
4   *Bavin* v *Bunney* (1950) 13 WN 181.
5   (1921), unreported, but extensively noted at [1931] 2 KB 317.
6   For a Scottish case in which an estate agent was held to have brought about a 'sale' on conclusion of missives, notwithstanding that there was no completion because of a suspensive condition, see *Chris Hart (Business Sales)* v *Currie* 1991 SLT 544.

The point of time at which a purchaser's willingness and ability must be proved to exist is normally the date fixed for completion, and the best method of proof is naturally to show that completion has taken place as arranged. However, where the vendor resiles from a binding contract and thus prevents completion from taking place, this form of evidence is not available to the agent, who may none the less claim commission. As Denning LJ explained:[1]

> The reason why the vendor is liable in such a case is because, once he repudiates the contract, the purchaser is no longer bound to do any more towards completion: and the vendor cannot rely on the non-completion in order to avoid payment of commission, for it is due to his own fault ... But if the vendor could show that the purchaser would not in any event have been able or willing to complete, he would not be liable for commission.

The requirement stated above, that the contract must be 'binding', was considered by the Court of Appeal in *Peter Long & Partners* v *Burns*.[2] The claimants there introduced a potential purchaser of the defendant's garage business and contracts were exchanged, but it was then discovered that the claimant's representative, acting on information received from the defendant, had innocently misrepresented the effect upon the property of a local Town and Country Planning scheme. On discovering the truth, the applicant sought rescission of the contract and was released from his obligations by the defendant. The Court of Appeal held that, as the present contract was voidable, it could not be described as 'binding' and the agents had not earned their commission.

The decision seems a rather harsh one, since the flaw in the contract had after all resulted from information provided by the client. However, it was followed by *Blake & Co* v *Sohn*,[3] where a contract of sale negotiated by the claimants was rescinded by the purchasers on the ground that the vendor could not prove title to the whole of the property. Nield J recognised that, if the failure of the sale was due to the vendor's 'default', the claimants would be entitled to their commission.[4] None the less, he held that the client's action, in instructing the claimants to sell property which he did not own, did

---

1   *Dennis Reed Ltd* v *Goody* [1950] 2 KB 277, 285.

2   [1956] 1 WLR 413, 1083. See also *Gregory* v *Fearn* [1953] 2 All ER 559.

3   [1969] 3 All ER 123. See also *C & S Realties of Ottowa* v *McCutcheon* (1978) 84 DLR (3d) 584.

4   See *Luxor (Eastbourne) Ltd* v *Cooper* [1941] AC 108: p 86.

not amount to 'default' for this purpose (no doubt the result would have been different if the client had known that his title was defective).

The judge in *John D Wood & Co (Residential and Agricultural) Ltd* v *Craze*,[1] while accepting the correctness of the principle laid down in Peter Long, explored the limits of that principle. The case arose after the claimant agents had introduced a purchaser for the defendant's flat. Contracts were exchanged and a deposit paid but the purchaser, upon discovering that the defendant had fraudulently provided false pre-contract information regarding existing disputes about the property, rescinded the contract. It was held that the agents were not entitled to commission, since their terms of business made this payable on the exchange of 'unconditional contracts of sale', and the contract in question could not be described as 'unconditional'. However, the judge went on to hold that the agents were entitled to damages for breach of a term necessarily implied into the agency agreement, that the vendor would not deliberately make false statements to a prospective purchaser and thus jeopardise the sale. The overall position thus appears to be that agents are protected against the client's fraud, but not against the client's innocent misrepresentations.

An important point, on which there appears to be no direct authority, is whether estate agent's entitlement to commission depends on the contract of sale being, not just legally binding, but specifically enforceable. In *Murdoch Lownie Ltd* v *Newman*[2] Slade J said:

> I think that 'binding contract' means a contract which binds both parties, ie, which is a contract and is clothed with the necessary form required to make in an enforceable contract in law. But whether that be so or not, I am satisfied that the expression envisages a contract which, as against the party introduced by the agent, the vendor is entitled to have specifically performed or for whose breach he is at least entitled to recover damages.

The view that, where specific performance would not be available, the estate agent does not qualify for commission, has been expressed on a number of occasions, notably by Denning LJ in *Fowler* v *Bratt*[3] and *McCallum* v *Hicks*.[4] However, in *Sheggia* v *Gradwell*,[5] commission was

---

1   [2008] 1 EGLR 17.
2   [1949] 2 All ER 783, 789.
3   [1950] 2 KB 96, 105.
4   [1950] 2 KB 271, 274. See also *Dennis Reed Ltd* v *Goody* [1950] 2 KB 277, 283, *per* Bucknill LJ.
5   [1963] 3 All ER 114. For the criticism which this decision has attracted, see p 95.

expressed to be payable in any of a number of events, one of which was: 'if within the said period of three months any person introduced by the agents enters into a legally binding contract to purchase the said business and property.' The agents introduced a person who signed a contract to purchase the property, which was leasehold; since, however, the landlord was not satisfied with this person's references, he was forced to break the contract. The Court of Appeal, by a majority (Lord Denning MR dissenting) held that this contract could properly be described as 'legally binding' notwithstanding that specific performance might not have been obtainable.

As we have seen, a vendor whose agent has procured a binding contract cannot, by breaking that contract, avoid paying for the agent's services. However, the vendor's obligation extends no further than that; where it is the purchaser who resiles from the binding contract, '[t]he vendor is not bound to bring an action for specific performance or for damages simply to enable the agent to get commission; but, if he does get his money, he will probably be liable to pay the commission out of it.'[1]

The first part of this suggestion came before the Court of Appeal in the case of *Boots* v *E Christopher & Co*,[2] where the defendant agents were to introduce 'a person who is willing and able to purchase' the claimant's shop in return for a commission 'at the rate of 5 per cent of the total purchase price obtained'. The defendants introduced a person who signed a contract to purchase and paid a deposit to the agents. Subsequently, however, this person, although financially able to complete the purchase, refused to do so and forfeited the deposit. The defendants, who claimed to have earned their commission, deducted this from the deposit and handed only the balance to the claimant. It was held that the words 'total purchase price obtained' meant that commission would not be earned until completion, since only then would the money be 'obtained' by the vendor. Further, the claimant was not at fault in merely forfeiting the deposit and choosing not to seek specific performance or damages. Thus the defendants were not entitled to commission and were ordered to pay over what they had wrongfully retained.

There appears to be no authority on the second point raised by Denning LJ, as to the liability of a vendor who successfully sues a recalcitrant purchaser. Clearly, if specific performance is awarded, the

---

1   *Dennis Reed Ltd* v *Goody* [1950] 2 KB 277, 285, *per* Denning LJ.
2   [1952] 1 KB 89.

agent is entitled to commission under the clause used, for completion will then take place.[1] So, too, if the vendor obtains an award of damages, it would seem equitable that the agent should receive something, although in this case it cannot be commission as such. In a case of this nature, it has been suggested[2] that the agent would succeed in an action for reasonable remuneration. This suggestion, although open to objection on technical grounds[3] would seem to do substantial justice.

The principles discussed above are undoubtedly regarded by the courts as the norm, so that any new clause is presumed to fall into the same category unless and until it is established that the language used indicates a clearly contrary intention. A good example of the prevailing judicial attitude is to be found in the speech of Lord Russell of Killowen in *Luxor (Eastbourne) Ltd* v *Cooper*:[4]

> It is possible that an owner may be willing to bind himself to pay a commission for the mere introduction of one who offers to purchase at the specified or minimum price, but such a construction of the contract would, in my opinion, require clear and unequivocal language.

Similar views have been expressed on a number of occasions by Lord Denning MR. Thus: 'in the absence of express terms to the contrary, the commission of the agents is to be paid out of the proceeds of sale'[5] and, likewise: 'The common understanding of mankind is that commission is only payable by the vendor when the property is sold.'[6] Indeed, Lord Denning has taken this principle a stage further:[7]

> When a house is in the hands of more than one agent, it cannot be supposed that the vendor is to be made liable for double commission. In

---

1 The argument that, even so, the purchaser is not 'willing' is unlikely to attract the sympathy of the courts.

2 By Denning LJ in *Boots* v *E Christopher & Co* [1952] 1 KB 89, 98. Indeed, Denning LJ suggested at p 99 that he would have made such an award; if only the agents had claimed on that basis.

3 It forces the client to pay for something not originally agreed between the parties (unless consent can be implied).

4 [1941] AC 108, 129.

5 *Fowler* v *Bratt* [1950] 2 KB 96, 104.

6 *Jaques* v *Lloyd D George & Partners Ltd* [1968] 2 All ER 187, 190. See also *McCallum* v *Hicks* [1950] 2 KB 271, 274 and *Dennis Reed Ltd* v *Goody* [1950] 2 KB 277, 284.

7 *AA Dickson & Co* v *O'Leary* [1980] 1 EGLR 25. Contrast the views expressed by Sir David Cairns at p 733, and by Orr LJ in *Christie Owen & Davies* v *Rapacioli* [1974] 1 QB 781, 790.

the ordinary way commission is payable to the agent who is the first to find a purchaser who enters into a binding contract which both parties accept. If a binding contract is made before another agent has produced 'a person able, ready and willing to purchase' — or before any contract is made with that person — the first agent gets commission, the second does not. It is a race as to which agent wins. He wins who first gets the binding contract.

Whether or not this view, that it is impossible for a vendor to become liable to more than one agent, can be supported is closely linked to the problem of whether or not commission can ever be recoverable in the absence of a contract between vendor and purchaser. This question is considered below.[1]

Exchange of contracts

The remarks of Lord Russell of Killowen quoted above have been said to have 'inspired estate agents to set about devising formulae intended to ensure that principals should not be able to escape liability by refusing to accept offers by persons willing to purchase'.[2] Before evaluating the results of this inspiration, however, we may consider another form of words which is designed to make the agent's right to commission rest solely upon the formation of a binding contract between vendor and purchaser, thus ruling out any discussion of the 'purchaser's' subsequent willingness or ability to complete the transaction. In *Midgley Estates Ltd* v *Hand*[3] the agreement between the agents and their client was contained in a letter from the agents which provided for the payment of commission as soon as 'our purchaser shall have signed a legally binding contract, effected within a period of three months from this date'. A person was introduced who signed a contract within the stipulated time, but this applicant proved financially unable to complete and so forfeited the deposit to the vendor. The Court of Appeal held that, as the terms of the agreement were clear and unambiguous, there was no reason for not enforcing them; accordingly the agents were entitled to their commission.

This decision is perhaps open to objection on the ground that the use of the word 'purchaser' should, in accordance with established

1 Pp 97–108.
2 *Graham & Scott (Southgate) Ltd* v *Oxlade* [1950] 2 KB 257, 262, *per* Cohen LJ.
3 [1952] 2 QB 432.

principles, have raised the question of that person's willingness and ability to complete. However, it was said by Jenkins LJ that the rest of the agreement so clearly qualified the word 'purchaser' that such questions were rendered irrelevant.

The *Midgley Estates* case was followed by a majority of the Court of Appeal in *Sheggia* v *Gradwell*,[1] and, while that decision has attracted a considerable amount of criticism,[2] more recent authority confirms that effect will be given to a clause which clearly ties the payment of commission to the exchange of contracts. This was the position in *Foxtons Ltd* v *Thesleff*,[3] where the claimants' terms of business provided, under the headings of both 'Sole Agency' and 'Multiple Agency', that commission would become payable: 'if at any time unconditional contracts for the sale of the property are exchanged with a purchaser ...'. In a separate clause, it was stated that: 'All Foxtons fees become payable upon exchange of contracts.' The agents introduced someone who, having exchanged contracts to purchase a client's property, later backed out of the contract and forfeited his deposit. The client sought to argue that the word 'purchaser' (as used in the Estate Agents (Provision of Information) Regulations 1991)[4] meant someone who actually completed the purchase, but this argument was unanimously rejected by the Court of Appeal on the ground that an agent's entitlement to commission is based, not on the 1991 Regulations, but on the agency contract itself.

Willing to purchase

The most important commission clauses to have been inspired by Lord Russell of Killowen's *dictum* are those which make commission payable 'on the introduction of a person willing and able to purchase' the property. This formula, with the addition or substitution of the words 'ready' and 'prepared', was much in vogue in the 1950s, its object being to prevent the vendor not only from breaking a binding contract, but also from refusing to enter into such a contract with a suitable prospective purchaser. Although, once again, the individuality

---

1   [1963] 2 All ER 114: p 93.
2   Lord Denning MR dissented. Further criticism came from Salmon LJ in *A L Wilkinson Ltd* v *Brown* [1966] 1 All ER 509, 515 and Edmund Davies LJ in *Jaques* v *Lloyd D George & Partners Ltd* [1968] 2 All ER 187, 192.
3   [2005] 2 EGLR 29.
4   See pp 129–130.

of each case must be stressed, the particular epithets attached to the potential purchaser have been interpreted sufficiently often by the courts for a number of principles to have emerged.

In *Graham & Scott (Southgate) Ltd* v *Oxlade*[1] the claimant estate agents introduced to the defendant vendor a person who made an offer for the property 'subject to contract' and 'subject to satisfactory survey'. This person was financially able to purchase the property and anxious to do so, the qualifications to her offer being added as a prudent precaution. The defendant, however, received a better offer from someone else, to whom the property was thereupon sold. The Court of Appeal held that the conditions attached to the first offer prevented the person making it from being described as 'willing' to purchase the property; accordingly the terms of the commission agreement were not fulfilled and the agents' claim failed.

The Court of Appeal in *Oxlade*'s case overruled the decision of Lewis J in *Giddy & Giddy* v *Horsfall*,[2] where a person introduced by the claimants was held to be 'willing' notwithstanding that his offer was made 'subject to contract'.[3] Two other decisions at first instance, those of Lynskey J in *Dennis Reed Ltd* v *Nicholls*[4] and Hilbery J in *E H Bennett & Partners* v *Millett*,[5] were also viewed with some suspicion. Nevertheless, it is at least doubtful whether they were actually overruled,[6] since it was not clear in either case whether the qualification 'subject to contract' was introduced by the prospective purchaser or by the vendor. If the vendor had accepted conditionally an offer which was unqualified, then, as was acknowledged by Cohen LJ in *Oxlade*'s case, the agent would be entitled to claim commission.[7]

The principle underlying *Oxlade*'s case was applied by the Court of Appeal to a different form of words in *Bennett, Walden & Co* v *Wood*.[8] The agents in that case wrote to the vendor confirming that 'in the

---

1   [1950] 2 KB 257.
2   [1947] 1 All ER 460.
3   The commission agreement in fact used the term 'prepared', but Lewis J treated this as synonymous with 'willing'.
4   [1948] 2 All ER 914.
5   [1949] 1 KB 362.
6   This was assumed to be so by Denning LJ in *Dennis Reed Ltd* v *Goody* [1950] 2 KB 277, 289.
7   This point was applied in the County Court case of *Lucas & Sons v Mayne* (1954) 164 EG 441.
8   [1950] 2 All ER 134.

event of our securing for you an offer of £2,900 or some such offer as shall be acceptable to you our commission will be at the recognised ... scale'. A person was introduced who made an offer 'subject to contract'. The vendor accepted this offer but changed his mind before contracts were exchanged, with the result that the sale did not go through. It was held that the agents were not entitled to commission, for the 'offer' which they were to obtain had to be a firm offer, not one made 'subject to contract'.

Where a person introduced by an estate agent actually enters into a binding contract to purchase property, that person's 'willingness' to do so is placed beyond dispute.[1] The problem inherent in the 'willing to purchase' cases, however, is whether, in the absence of such a contract, the estate agent may establish an applicant's willingness by means of alternative evidence. This question has received a remarkable divergence of judicial answers, and the two main views expressed must now be examined.

The leading proponent of what may be called the 'narrow view', that is, the view that no other evidence will suffice, is Lord Denning MR, whose determination that clients should never be made to pay more than one commission upon a single property has led him to distrust any clause which awards the agent commission on less than a completed sale. The opinion of the learned Master of the Rolls is effectively illustrated by his assertion in *McCallum* v *Hicks*[2] that:

> A person may not properly be said to be 'willing' to purchase, so as to entitle an agent to commission, unless he is irrevocably willing, that is, unless he has given irrevocable proof of his willingness by entering into a binding contract to purchase.

If correct, 'this interpretation means that the special clause has practically the same effect as the usual terms on which an estate agent is employed'[3] thus making a binding contract and continued willingness to complete prerequisites of the agent's claim. Support for this view is to be found in *Martin, Gale & Wright* v *Buswell*[4] where

---

1   The argument that 'willingness' and 'ability' are to be judged at the time of introduction, rather than at the point of sale, was rejected in *Knight Frank & Rutley* v *Fraser* 1974 SLT 50.

2   [1950] 2 KB 271, 276; see also *Dennis Reed Ltd* v *Goody* [1950] 2 KB 277, 288.

3   *Dennis Reed Ltd* v *Goody* [1950] 2 KB 277, 287, *per* Denning LJ.

4   (1961) 178 EG 709. See also *Mustafa* v *K G Palos* (1972) 224 EG 35.

commission was expressed to be payable on the introduction of a person 'prepared to purchase on the terms of your instructions, or on terms acceptable to you'. After an initial offer made 'subject to contract', a price was agreed and the parties instructed their solicitors. The prospective purchaser's solicitor, having received a draft contract, returned it marked 'approved' and sent an engrossment to his client for signature. Before this was signed, however, the vendor withdrew from the sale. Upon these facts the Court of Appeal rejected the estate agent's claim for commission, and there are undoubtedly passages in the judgments which suggest that it was the absence of a binding contract which prevented the applicant from being described as 'prepared'. However, the case may also be explained upon the ground that the applicant's original offer remained 'subject to contract' until such time as it could clearly be shown to have become unconditional (presumably when he returned the signed contract to his solicitor).

The 'broad view', which permits an estate agent to prove an applicant's willingness merely by showing that the latter has made an unqualified offer to purchase, was clearly taken by Bucknill LJ in *Dennis Reed Ltd* v *Goody*.[1] Although the agents' claim in that case failed, because their applicant withdrew from negotiations before contracts were exchanged, Bucknill LJ stated that they could have succeeded by showing 'that he was ready, able and willing to purchase up to the time when either an enforceable contract for the purchase of the house is made between the parties, or, alternatively, up to the time when the vendor refuses to enter into such a contract on terms on which the purchaser is willing to purchase and the vendor was at one time willing to sell'.

The opinion of Bucknill LJ was clearly *obiter*, as were the similar views expressed by the Court of Appeal in *E P Nelson & Co* v *Rolfe*.[2] The court there was satisfied as to the 'willingness' of a person who never in fact made an offer, even conditionally, for the property, but the point in that case was conceded by counsel for the client. However, three later cases provide direct authority for the broad view.

In *John E Trinder & Partners* v *Haggis*,[3] the claimants introduced a person who signed his part of a contract to purchase the property; when the vendor refused to exchange contracts, the Court of Appeal

---

1   [1950] 2 KB 277, 283.
2   [1950] 1 KB 139.
3   (1951) 158 EG 4.

held by a majority that he was liable to pay commission. A more extreme case is *AL Wilkinson Ltd v O'Neil*,[1] in which commission was made payable upon the introduction of 'an applicant willing and able to sign a contract to purchase' at a named price or such other price as the vendor would accept. Negotiations reached the point at which the prospective purchaser received a draft contract; this he returned with a number of small verbal amendments, and one of more substance. The Court of Appeal accepted evidence that the applicant would have waived this amendment if pressed to do so and, stressing that 'willingness' is a matter of fact, found it to be satisfactorily established in this case.

The third, and most important, case is *Christie Owen & Davies v Rapacioli*[2] in which, to earn commission, the agents were to effect 'an introduction either directly or indirectly of a person ready able and willing to purchase' the defendant's restaurant for £20,000 or an acceptable lower price. The claimants introduced a person who, having agreed a price with the vendor, signed his part of the contract and paid a deposit. The Court of Appeal, after considering the cases already referred to, decided that this person clearly fitted the description and therefore awarded the claimants their commission.

The weight of authority thus appears to favour the broad view[3] and, at least at first sight, it seems perfectly reasonable that this should be so. As Ormerod LJ said in *Ackroyd & Sons v Hasan*:[4] 'I can find no justification for holding that the words 'prepared to enter into a contract' mean the same as, 'introduction of a party who does enter into a contract'.' However, there is one argument in favour of the narrow view, put forward by writers on this subject,[5] which is extremely difficult to answer. The main lines of this argument are as follows: an applicant's willingness to purchase is only legally relevant if the terms upon which he or she is willing are acceptable to the vendor: the vendor's state of mind can seldom be ascertained before contracts are exchanged, because until that time the vendor is free (at least as against the applicant) to reject the offer outright or modify the terms for

---

1   (1961) 181 EG 137.

2   [1974] 2 All ER 311.

3   Inferential support may be also found in the drafting of the Estate Agents (Provision of Information) Regulations 1991: see p 105.

4   [1960] 2 QB 144, 163.

5   Ash, *Willing to Purchase*, Heriot Press, 1963; *Bowstead & Reynolds on Agency*, 17th ed, 7–019.

acceptance: the commission agreement should not be interpreted in such a way as to fetter this freedom, for 'the commission agreement is, however, subordinate to the hoped for principal agreement of sale. It would be strange if what was preliminary or accessory should control the freedom of action of the principal in regard to the main transaction which everyone contemplates might never materialise.'[1]

This argument does not appear to have been specifically canvassed in the courts; it is significant, however, that in *Ackroyd & Sons* v *Hasan*,[2] where the commission agreement expressly required the vendor's assent to the terms of purchase, the Court of Appeal held this to be lacking despite the fact that the parties' solicitors had reached agreement. If, then, the vendor's assent is impliedly required in the normal 'willing to purchase' case, it is difficult to resist the conclusion that this requires an exchange of contracts, or at least the signature by the parties of their respective parts.

The pronouncements of the Court of Appeal in *AA Dickson & Co* v *O'Leary*[3] did little to dispel the general confusion surrounding the question of 'willingness' to purchase. The defendant in that case, who wished to sell his flat, put it in the hands of more than one firm of agents. The claimants, who were to receive commission on 'either directly or indirectly introducing a person ready, able and willing to purchase on terms authorised by [the defendant]', introduced an applicant who seriously intended to purchase and who, on 28 October, actually signed her part of a contract and returned it to her solicitors. On receiving this document on 29 October, the solicitors immediately telephoned the defendants' solicitors, only to be told that contracts had been exchanged on 28 October with a rival applicant. The claimants nevertheless claimed to be entitled to their commission, but it was held by the Court of Appeal[4] that commission is not payable unless the agent carries out his instructions before the property is sold to someone else (whether or not the agent is aware of that fact).

As to precisely when the claimants' applicant might have been described as 'willing', Lord Denning MR thought that this would be when the part of the contract which she had signed reached her solicitors (why this should be preferred to the moment of signature was not made clear). Sir David Cairns, however, in a discussion with counsel

---

1   *Luxor (Eastbourne) Ltd* v *Cooper* [1941] AC 108, 138–9, *per* Lord Wright.
2   [1960] 2 QB 144.
3   [1980] 1 EGLR 25.
4   Following *E P Nelson Co* v *Rolfe* [1950] 1 KB 139.

on costs which is not contained in the published report of the case, suggested that a person's willingness would not be established until the transaction was beyond recall, which would be when that person's solicitors actually *sent* the signed part of the contract to the other side.

For the moment at least, the last word on this subject lies with the Court of Appeal in *Savills Land & Property Ltd* v *Kibble*,[1] a decision which gives considerable encouragement to estate agents to use 'ready willing and able' as a basis for the payment of commission. The case in fact concerned, not commission as such, but an attempt by the claimants to provide for a 'withdrawal fee' in the event that a client refused to go through with a proposed sale of agricultural property. This was done by means of clause 5 in the claimants' standard terms of business, which provided that: 'the client will be liable to pay half the agreed sale fee to [the claimants], in addition to any other costs or charges agreed, if a ready, willing and able purchaser is introduced to the property and terms are agreed for the sale in accordance with the client's instructions, while [the claimants] have sole selling rights or joint sole selling rights, and this must be paid if the client subsequently withdraws and unconditional contracts for sale are not exchanged.'

The claimants introduced a prospective purchaser who offered a price acceptable to the client, and the parties reached agreement 'subject to contract'. However, the client then sought substantially to alter the terms of that agreement and, when the purchasers were not willing to accept the alterations, the sale collapsed. In interpreting clause 5, the Court of Appeal was in no doubt that the applicants introduced by the claimants satisfied the definition of 'ready, willing and able purchasers', notwithstanding that the parties had never come close to exchanging contracts. Moreover, terms had indeed been 'agreed' (albeit subject to contract) in accordance with the client's instructions, and the client must be treated as having withdrawn. It followed that the claimants were entitled to their fee.

Ready, able and prepared

Whether or not a person is 'able' to purchase property is regarded by the courts as depending primarily upon questions of finance. In this connection the following words of Atkin LJ[2] have often been quoted:

---

1   [1998] EGCS 170.
2   *James* v *Smith* [1931] 2 KB 317n, 322.

I think that 'ability' does not depend upon whether the purchaser has got the money in hand at the time or the balance at his bank. To my mind it is a question of fact. I do not think it depends on whether he has a binding agreement by which some third person is obliged to provide him with resources to carry out the contract. I think it is sufficient if it is proved by the agent or by the purchaser the circumstances are such that if the vendor had been willing and ready to carry out his contract he on his part at the proper time could have found the necessary money to perform his obligation.

This *dictum* was applied by Lynskey J in *Dennis Reed Ltd* v *Nicholls*[1] where, after hearing evidence as to the potential purchaser's earnings and the value of the house which he owned, the learned judge found that he was clearly able to purchase the property in question.

'Ability' was extended to include matters other than finance in the case of *Dellafiora* v *Lester; Lester* v *Adrian Barr & Co Ltd*.[2] In that case the agents, in return for a commission of 10 per cent of the purchase price, were instructed to find a person willing and able to purchase property which consisted of the remaining 11 years of a lease on a café. The agents introduced a person who signed a contract and paid a deposit. However, the sale went off because the landlords refused their consent to the assignment, and the purchaser's deposit was returned in accordance with the conditions of sale. The agents, who earned commission on a subsequent sale of the property, claimed to be entitled in addition to commission on this abortive introduction. The Court of Appeal held that the person whom the agents introduced could not be described as 'able' in the face of the landlord's refusal to allow her to purchase the property; accordingly the agents' claim failed.

There is little authority upon the meaning, if any, which is to be given to the word 'ready' where this is used in addition to, or in substitution for, 'willing and able'. The view of Denning LJ that a person who is ready 'must have made all necessary preparations by having the cash or a banker's draft ready to hand over'[3] does not appear to be based on previous decisions; nor has it been followed. It is submitted that 'ready' adds nothing to the requirement already discussed, a view which is supported, for example, by the judgment of Atkin LJ in *James* v *Smith*[4] where the terms 'ready' and 'able' are used interchangeably.

---

1   [1948] 2 All ER 914.
2   [1962] 3 All ER 393.
3   *Dennis Reed Ltd* v *Goody* [1950] 2 KB 277, 287.
4   [1931] 2 KB 317n, 322.

As a variation upon the 'willing and able' theme, some estate agents stipulate that commission shall be payable on the introduction of a person 'prepared to enter into a contract' to purchase. The Court of Appeal in *Ackroyd & Sons* v *Hasan*[1] had no difficulty in finding that the prospective purchasers were 'prepared', since they had gone so far as to sign their part of the contract. No general consideration of the meaning of 'prepared' took place, however, and for guidance on this matter one must turn to the later case of *A L Wilkinson Ltd* v *Brown*[2] which concerned the sale of a leasehold property. The claimant estate agents introduced a person who was undoubtedly eager to purchase, but whose references at first failed to satisfy the landlord. This obstacle was later removed, but the potential purchaser could not exchange contracts until he was sure of selling his own property and raising the purchase price. Before his expectations were fulfilled, the defendant sold the property elsewhere. The Court of Appeal, in holding that the agents were not entitled to commission because the person they had introduced was never 'prepared' to purchase, considered the meaning of that term. Harman LJ said:

> '[P]repared', as I think is conceded, means something more, generally speaking, than 'willing'. It means a person who is ready and willing. It may involved the qualification 'able', but in many circumstances one would not look beyond whether the person was a genuine person ready and willing to sign a contract.

The meaning of this remark is somewhat obscured by the use of the word 'ready' which, as we have already seen, means 'able' if it means anything at all. Salmon LJ, on the other hand, would not concede that the agent's position was in any way improved by the use of the word 'prepared'. 'In my judgment a person is not prepared to enter into a contract who is not ready, willing and able to do so.'

Willing purchaser

The commission clauses so far considered draw a clear distinction between a 'purchaser' and a 'person willing to purchase'. Since, however, it is common to find estate agents using commission terms

---

1   [1960] 2 QB 144.

2   [1966] 1 All ER 509.

which envisage the introduction of a 'willing purchaser', it is relevant to consider into which category such a clause falls. As we have already seen, a 'purchaser' is treated by the courts as meaning one who actually purchases, and it is therefore strongly arguable that 'willing purchaser' cannot mean less than this. If it means anything at all, it surely suggests that the purchaser must be shown to have purchased willingly. As against this, there are at least two decisions in which the normal meaning of 'purchaser' was held to have been qualified by the context in which it was found. In *E H Bennett & Partners* v *Millett*,[1] a clause requiring an agent to introduce 'a purchaser who is able and willing to complete the transaction' was held to have been satisfied by the introduction of someone who never in fact signed a contract but who remained at all times willing and able to do so. And in *Midgley Estates Ltd* v *Hand*,[2] the characteristics usually required of a 'purchaser' (continued willingness and ability to complete the purchase) were held to have been rendered irrelevant by making commission payable as soon as 'our purchaser shall have signed a legally binding contract'.

In so far as these decisions suggest that the meaning of 'purchaser' may vary, they lend support to the view that 'willing purchaser' is to be equated with 'person willing to purchase'. However, in what appears to be the only reported case in which the phrase itself required interpretation, the Court of Appeal took a much stricter line. The case in question, *Davis* v *George Trollope & Sons*,[3] in fact concerned, not a straightforward commission clause, but an agreement between an agent and a sub-agent. The claimant, an estate agent, thought that he could find a purchaser for a property which was on the defendants' books, and the defendants agreed to pay him one-half of the sum which they would earn if he introduced to them a willing buyer at an agreed price. The claimant made his introduction, but the vendor sold the house elsewhere and the defendants did not therefore earn any commission. This did not prevent the claimant from claiming against the defendants and, in the county court, the claim succeeded. The Court of Appeal, however, reversed this decision on two grounds: first, the term 'willing buyer' meant, not 'person willing to buy' but 'person who actually becomes a buyer', something which had clearly not happened; second, in the view of Scott LJ, a contract of this nature must

---

1   [1949] 1 KB 362.
2   [1952] 2 QB 432.
3   [1943] 1 All ER 501.

mean: 'If through your bringing a willing buyer I earn my commission, I will then pay you half of it.'

Although the true meaning of 'willing purchaser' is open to considerable doubt, it seems that the issue has been settled for practical purposes by the Estate Agents (Provision of Information) Regulations 1991. Those regulations require any estate agent using the term 'ready, willing and able purchaser', or an equivalent, to explain to the client in writing what the term means, using the forms of explanation contained in the regulations themselves. The definition given is a person who 'is prepared and is able to exchange unconditional contracts for the purchase of your property', and the statutory wording required continues:

> You will be liable to pay remuneration to us, in addition to any other costs or charges agreed, if such a purchaser is introduced by us in accordance with your instructions and this must be paid even if you subsequently withdraw and unconditional contracts for sale are not exchanged, irrespective of your reasons.

The effect of these provisions is to equate 'ready, willing able purchaser' with 'person ready, willing and able to purchase', the meaning of which was considered earlier.[1] Thus, agents who fulfil their statutory obligations by incorporating the official definition into their commission agreements will have the benefit of the wider interpretation and will be entitled to payment, *Davis* v *Trollope* notwithstanding.

Other clauses

From all that has gone before, it is clear that an estate agent may stipulate for the payment of commission notwithstanding that no contract has been made, (eg the 'willing and able' cases) and notwithstanding the inability of a purchaser to complete a contract which has been made (where commission is linked merely to the contract). It remains for us now to consider two attempts which have been made to provide for the payment of commission without reference to either of these conditions. In *Drewery* v *Ware-Lane*[2] the defendant, who owned a leasehold house, put it into the hands of the claimants

---

1   See pp 97–105.
2   [1960] 3 All ER 529.

and signed an agreement to pay them commission 'if and when (a) a prospective purchaser signs your 'purchaser's agreement' and (b) I sign your 'vendor's agreement' (receipt of a copy of a form of each of the above agreements is hereby acknowledged)'. The agents introduced a person, a price was agreed, and the two 'agreements' were signed 'subject to contract'. The person introduced was not, at that stage, able to purchase the property, as he had first to obtain a mortgage; however, since he wished to buy the freehold from the landlord, he did not intend to apply for a mortgage until his negotiations with the landlord had terminated, successfully or otherwise. Nine days after the signing of the agreements, the defendant sold the property to someone else and the agents sued for their commission. The Court of Appeal held that the terms of the commission agreement were clear and had been fulfilled by the agents. In considering whether the applicant could properly be described as a 'prospective purchaser', Ormerod LJ said:

> The word 'prospective' does not connote necessarily either the term 'ready' or 'willing' or 'able'; it means a man who has the question of buying this property in prospect or in contemplation and is prepared to make an offer with regard to it.

The second attempt by estate agents to cross the usual barriers met with a good deal less success. In *Jaques* v *Lloyd D George & Partners Ltd*[1] the claimant agreed that he would pay commission to agents 'should you be instrumental in introducing a person willing to sign a document capable of becoming a contract to purchase at a price, which at any stage of the negotiations has been agreed by me'. As we shall see,[2] the agents were not allowed to rely on this clause since they had misled the claimant as to its effect. However, Lord Denning MR and Cairns J provided a further ground for the decision, holding the clause unenforceable by reason of its vagueness and uncertainty. As Cairns J pointed out, one possible reading of this clause would entitle the agents to commission if the person introduced signed a blank piece of paper, since this could then be completed so as to form a contract of sale!

---

1   [1968] 2 All ER 187.
2   P 140.

Auctions — commission on sale

As in the case of estate agents, whatever contract has been made between auctioneer and client must be carefully construed, in order to establish exactly what is required to happen before commission becomes payable. In *Peacock* v *Freeman*[1] the claimants asked the defendants, a firm of auctioneers, what commission they would require for selling certain property. After some dispute a scale of charges was settled which would apply in the event of a sale and it was further agreed that, if the property were not sold, a fee of thirty guineas together with out-of-pocket expenses would be payable. The property was duly knocked down at the sale, and a deposit paid to the auctioneers. Later, however, the purchaser objected to the title and the vendors, in accordance with the conditions of sale, rescinded the contract and repaid the purchaser the amount of the deposit. The vendors then sought to recover the deposit money from the auctioneers; they, however, claimed to deduct from it their full commission, as the property had been 'sold'. The Court of Appeal held that the auctioneers were entitled only to the agreed fee of thirty guineas and expenses, as the commission agreement contemplated a completed sale and not merely the fall of the hammer. Further, since the vendors had acted in accordance with the conditions of sale in rescinding the contract, they could not be said to have wrongfully deprived the auctioneers of their commission.

It must be emphasised that the decision in *Peacock* v *Freeman* was based, not on any rule of law, but on the interpretation of the agreement made by the parties. This point is illustrated by the case of *Skinner* v *Andrews & Hall*[2] where the terms upon which the auctioneers were to be paid commission were contained in letters to and from their client. These stipulated that half-commission would be payable if the property, though not sold at the auction, were sold within two months thereafter to a person introduced by the auctioneers. It was also stated that commission would be payable on any sale taking place before the sale under the hammer or before 30 October but that, if the property remained unsold at that date, no charge of any description would be made. The property was knocked down at the auction but the vendor then rescinded the contract, as he was entitled to do under

1   (1888) 4 TLR 541.
2   (1910) 26 TLR 340.

the conditions of sale. It was held nevertheless that the auctioneers had earned their commission, for it was clear that in this case the parties had used the word 'sale' to mean the knocking down of the property to a purchaser, rather than the completion of the transaction by a formal conveyance.

Auctions — sale by private treaty

It sometimes happens that, after an abortive auction, the property is purchased privately by a person who either has been introduced to the vendor by the auctioneer or who has at least learned about the property from advertisements of the sale. In such a case the auctioneer may be entitled to commission on the sale but, once again, this depends upon the terms of the agreement with the client. What is certain is that an auctioneer who is not authorised to sell by private treaty cannot earn commission by purporting to sign a contract on behalf of the vendor. Thus in *Marsh* v *Jelf*,[1] an auctioneer who negotiated a private sale of the property was not allowed to bring evidence of a custom among auctioneers that commission should be payable; as Keating J said: 'Auctioneers could not among themselves make such a custom to bind the rest of her Majesty's subjects.'

Where an auctioneer *is* authorised to sell by private treaty, any claim to commission will depend on the same principles as those which govern the claims of estate agents. In *Green* v *Bartlett*,[2] for example, the claimant was instructed to sell an island for the defendant 'by public auction or otherwise'. The claimant was held to be entitled to commission upon a private sale which was proved to have been negotiated by him, although the sale was concluded by the defendant personally.

The commission clause which came before the House of Lords in *Bayley* v *Chadwick*[3] was of an unusually wide nature, in that it provided for payment to the auctioneer upon a private sale to any person 'led to make an offer in consequence of your mention or publication for auction purposes'. The purchaser never met the auctioneer, nor did he see any advertisement of the sale; he was merely told by a third party,

---

1   (1862) 3 F & F 234; see p 168.
2   (1863) 14 CBNS 681.
3   (1878) 39 LT 429.

who attended the sale, that the property (a ship) had not been sold. The House of Lords, reversing the decision of the Court of Appeal and restoring that of the Court of Common Pleas, held that the jury were quite entitled to find that this purchaser came within the clause, so as to entitle the auctioneer to commission.

Substantial performance

In this chapter we have so far tended to assume that, once the event on which commission is payable has been satisfactorily identified, the question whether or not it has taken place is a simple and straightforward one. In practice, however, cases frequently arise in which what has actually occurred is not the stipulated event but something which resembles it to a greater or lesser degree. The problem which confronts the court in such a case is to decide whether the agent's entitlement to commission depends upon the exact fulfilment of the client's instructions or whether substantial performance will suffice. If the latter applies, the court must then decide whether what has been produced is sufficiently akin to what was envisaged for it to be said that the agent's instructions have been substantially performed.

A simple example of this principle in operation, and one with which every practising estate agent will be familiar, is that of a sale of property which goes through only upon the agreement of the vendor to reduce the asking price. In *Jack Windle Ltd* v *Brierley*[1] the claimants introduced a prospective purchaser for the defendant's bakery business, but he was unable to raise the purchase money and negotiations ceased. A short time later, after the claimants' instructions had been withdrawn, the defendant agreed to reduce the price to this applicant and to allow a part of the money to remain outstanding on the security of a second mortgage. It was held that the claimants were not entitled to commission, for they had not fulfilled their instructions; the sale which actually took place was effectively brought about by the vendor himself.

This, it must be said, seems a very harsh decision and one, moreover, which is somewhat out of line with previous authority.[2] Indeed, it has been criticised in subsequent cases,[3] and it now seems

---

1   [1952] 1 All ER 398. Like many such cases, this might equally well be analysed in terms of 'effective cause': p 114.

2   *Price Davies & Co* v *Smith* (1929) 45 TLR 542.

3   *Levers* v *Dunsdon* (1967) 206 EG 979; *Glentree Estates Ltd* v *Gee* [1981] 2 EGLR 28.

safe to assume that a price reduction in itself will not affect an agent's right to claim commission for an effective introduction.

A second group of cases raising the question of substantial performance concerns limited companies. Where, for example, the property which is to be sold represents virtually the only assets of a company, it sometimes happens that a person introduced by an estate agent buys, not the property, but the company itself, by acquiring all (or nearly all) its shares.

While the merits of such a situation appear to lie with the agent, to award commission causes substantial legal problems, not least that this requires the court to treat the seller(s) of the shares as the client.[1] Nevertheless, and, despite one decision to the contrary[2] the majority view is in the agent's favour.[3] However, an agent whose client is a company may still be worse off in one situation, namely, where the property is subject to a mortgage. It has always been accepted that commission on the sale of mortgaged property is based, not merely on the price paid for the equity, but on the 'grossed up' value. Where, however, mortgaged property is put on the market by a company, and an applicant introduced by the agent then buys the company's shares at a price which is depressed because of the mortgage, it has been held that the agent's commission can only be based upon the actual price of the shares; here no 'grossing up' is possible.[4]

Problems can also arise where an agent introduces a company as a potential purchaser of the client's property, for the labyrinthine ways of the business world may well mean that the property is ultimately transferred, not to the company actually introduced, but to another company which may be its subsidiary or linked to it in some other way. In the Australian case of *LJ Hooker Ltd* v *WJ Adams Estates Pty Ltd*,[5] for example, a company introduced by the claimants made an agreement

---

1   This is because the company itself is a legal person quite separate from its shareholders, whose identity is a matter of indifference to the company. It would hardly pay an agent to bring about a transfer of its shares from one person to another.

2   *Harris & Gillow* v *Kelly* (1953) 162 EG 622.

3   *Levers* v *Dunsdon* (1967) 206 EG 979; *Allen* v *Anderson* [1969] NZLR 951. In *Freedman* v *Union Group Ltd* [1997] EGCS 28, the agreement of the managing director and major shareholder to pay commission on the 'anticipated transaction' was held to cover a sale of the company's entire share capital.

4   *Way & Waller Ltd* v *Ryde* [1944] 1 All ER 9.

5   (1977) 138 CLR 52.

with another company which had already been negotiating with the vendor on terms that, no matter which of them ultimately purchased the site, they would join together in its development. When the land was actually sold to the other company, which then allotted shares to the claimants' applicant, it was held that the claimants were not entitled to any commission from the vendor; they could not be said to have caused the sale, nor could their applicant be described as a 'purchaser'. However every case turns on its own facts, so that, in another Australian case[1] concerning the sale of a leasehold farm, commission was held to be payable when the vendor and purchaser set up a joint company to manage the property for a period. Similarly, in *Gunn* v *Showell's Brewery Co Ltd*,[2] where the claimant was retained by the defendants to find a suitable brewery for them to purchase, it was held that they could not avoid the payment of commission by arranging for the property to be bought by one of their subsidiary companies.[3]

Apart from these identifiable groups, variations on the theme of substantial performance are endless. In *Griffin & Son* v *Cheesewright*,[4] where agents were instructed to let a house for three years, the applicant whom they introduced decided instead to purchase the vendor's entire interest in the property. In *Battams* v *Tompkins*,[5] the terms of a sale effected by the claimant were materially different from those envisaged by the client's instructions. In *Mason* v *Clifton (Bart)*,[6] it was the terms of the stipulated loan which differed. In all these cases, the agents failed in their claims for commission.[7]

In *G T Hodges & Sons* v *Hackbridge Park Residential Hotel Ltd*,[8] the claimants, in attempting to find a purchaser for the defendants' hotel, introduced a representative of the War Department which, however, was not prepared to pay the price which the defendants wanted. The

---

1   *Lord* v *Trippe* (1977) 51 ALJR 574.
2   (1902) 18 TLR 659. See also *Druce Investments Ltd* v *Thaker* [1988] 2 EGLR 229; *Kinney & Green* v *Johns* [1985] 2 EGLR 46.
3   In *Peter Yates & Co* v *Bullock* [1990] 2 EGLR 24, estate agents successfully claimed commission when an applicant whom they had introduced purchased the property jointly with another party.
4   (1885) 2 TLR 99.
5   (1892) 8 TLR 707.
6   (1863) 3 F & F 899.
7   See also *Dowling Kerr Ltd* v *Scott* [1996] EGCS 177, where an agent given 'sole selling rights' to procure either a freehold sale or a 21-year lease of property was held not entitled to commission when the client himself granted a 12-year lease.
8   [1940] 1 KB 404.

following year, as a result of their representative's report, the War Department acquired the hotel by compulsory purchase at a price even lower than their original offer. The Court of Appeal held that, in the absence of a voluntary sale, there was no contractual relationship between the agents' acts and the transfer of the property. The claimants were not entitled to commission, for all they had done was 'to start a train of causes which ultimately led to the defendants' property being taken away from them against their will',[1] at a price considerably less than they were willing to accept.

While most agents would probably regard the decision in the *Hodges* case as entirely reasonable, the same can hardly be said of the decision of the Court of Appeal in *Spiers (trading as 'Garmans')* v *Taylor*.[2] The claimants there, on being instructed to market the defendant's house, stated that the it would be 'entered on the Agent's list of properties for sale at the asking price of £34,500'. Commission would then be payable, it was said, on a sale at or above that price, or at any price subsequently agreed. For reasons which are not at all clear, the claimants never marketed the property at £34,500; they told the first serious applicant to appear that he need offer no more than £33,500 (while telling the client that a sale had been agreed at the asking price) and the client, in order to avoid breaking a chain of sales, agreed to sell at the lower price. The Court of Appeal held (quite rightly, it is submitted) that the agents must pay damages for depriving their client of the chance of obtaining a higher price. The court also held (quite wrongly, it is submitted) that the essential term which must be satisfied in order for commission to be earned was the marketing of the property at the correct price; since this had not occurred, no commission was payable, even though the house had been sold at (presumably) a fair price.

Further examples of transactions too far removed from the original instructions to entitle an estate agent to commission may be found in *Barnett* v *Isaacson*,[3] where a business was bought by someone whom the agent had introduced as a sub-agent, and *Richard Ellis* v *Pipe-Chem (Holdings) Ltd*[4] where agents acting for prospective tenants of business premises were only able to obtain an underlease, rather

---

1   At p 412, *per* Clauson LJ.
2   [1984] 2 EGLR 34.
3   (1888) 4 TLR 645.
4   [1981] 1 EGLR 21.

than the specified lease, since the landlords refused to lease the property directly to the agents' clients. Finally, as a monument to the eternal optimism of those who work on commission terms, we may look at the case of *Tufnell* v *Richard Costain Ltd*.[1] The claimant there, who was commissioned to procure finance for a large development project contemplated by the defendants' client, found himself unable to obtain the necessary loans. The claimant thereupon persuaded the client to use his own resources and duly claimed his commission for so doing! Not surprisingly, the claim failed, Paul J holding that the availability of this source of finance was due to the client's own decision rather than to the efforts of the agent.

The foregoing selection of unsuccessful commission claims serves to indicate that an agent who seeks to be paid for producing something different from what was originally called for faces something of an uphill struggle. However, the task is not an impossible one. In *Rimmer* v *Knowles*,[2] for example, an agent instructed to find a purchaser was held entitled to commission for introducing someone who took a 99-year lease of the property with an option to purchase the reversion. In *Hegarty* v *Bryant Homes Ltd*,[3] an agent seeking development sites earned commission for an introduction, even though the client then purchased the property by tender.[4] Again, in *Connells Estate Agents* v *Begej*,[5] an agent employed to bring about a 'sale' of the client's property recovered commission on a part-exchange; the amount of commission, moreover, was assessed on the full value of the transaction and not merely on the cash element.[6]

From an agent's point of view, the most generous decision is surely that of the Court of Appeal in *Chamberlain & Willows* v *HBS (Trust) Ltd*.[7] The claimants there introduced a prospective tenant for their clients' property but, before a lease could be signed, the clients sold the freehold. When the claimants' applicant subsequently leased

---

1   (1968) 209 EG 705.

2   (1874) 30 LT 496.

3   (1983, unreported).

4   In *Taplin* v *Barrett* (1889) 6 TLR 30, by contrast, no commission was payable by a vendor where the agent introduced someone who purchased the property at a subsequent auction.

5   [1993] 2 EGLR 35.

6   According to the Court of Appeal, a pure exchange of properties would not have satisfied the requirement of a 'sale'.

7   (1962) 184 EG 849.

the property from the new owners, the Court of Appeal[1] held that the agents could be said to have 'introduced a tenant' in accordance with their instructions; the clients were therefore liable to pay commission.

Effective cause

In order for an estate agent to claim commission it must be shown, not only that a specified event has occurred, but also that the agent has caused it to occur.[2] In 'willing to purchase' cases, the agent must have caused the applicant's 'willingness'.[3] This, like all questions of causation, is a matter of fact which is to be determined upon the evidence available. Moreover, it is a value judgment as to whether what the agent has done is sufficiently closely connected with the subsequent transaction for it to be said effectively to have caused the transaction, wholly or in part.

In this connection it should be noted that it is not enough for the agent's activities to be a *causa sine qua non*, that is to say, that the event would not have taken place without the agent's intervention (for example because the parties would never have met). It must further be established that the agent is an 'effective' or 'efficient' cause or, to put it another way, that the result has been achieved through the instrumentality of the agent. All these forms of words have been used in cases, by judges seeking to explain the legal position. The basic rule may be better understood, however, in the light of a hypothetical example. Let us suppose that an estate agent, who has been instructed to find either a tenant or a purchaser of a client's property, introduces a tenant, and that this tenant, many years later, purchases the freehold. The agent's act is clearly a *causa sine qua non* of the purchase since, in the final analysis, there would have been no purchase without the original introduction. However, common sense dictates that, for practical purposes, the agent cannot be treated as having 'caused' the sale to take place and so the agent will not receive commission on it. If, however, in similar circumstances, a person introduced by the estate agent as a tenant agreed in a matter of days to become a purchaser, the agent might well be treated as an 'effective cause' and thus entitled to commission.

---

1   Rather surprisingly, perhaps, reversing the trial judge (183 EG 417).
2   See Murdoch: 'The Principle of Effective Cause' (1985) 276 EG 742, 877.
3   *Hartnell, Taylor, Cook* v *Bromwich* [1982] 2 EGLR 32.

Nature and rationale

The effective cause principle has featured in many estate agents' commission disputes over the last century or so. The results of this voluminous case law were succinctly summarised by Lord Neuberger, sitting in the Court of Appeal in *Foxtons Ltd* v *Pelkey Bicknell*:[1]

- A term making an agent's commission dependent upon the agent being the effective cause of the relevant transaction will be 'very readily' implied into an estate agency agreement, especially in a residential consumer context, unless the provisions of a particular contract or the facts of a particular case negative it.
- The main reason for implying the term is to minimise the risk of a seller having to pay two commissions.
- It is not entirely clear whether the test is 'an effective cause' or 'the effective cause'.
- Whether an agent was the effective cause is a question whose resolution turns very much on the facts of the particular case.
- While two commissions are to be avoided, there will be cases where the terms of the relevant contracts and the facts compel such a result.
- Where the term is implied, the burden is on the agent seeking the commission to establish that he or she was the effective cause.

According to this passage in Lord Neuberger's judgment (and also to the vast majority of the case law), the effective cause principle is based on an implied term in the agency agreement. However, there is an alternative approach, according to which the 'effective cause' requirement derives from an interpretation of the actual words of the agreement, notably where a phrase such as 'introduce a purchaser' is used. This formulation appears to have its origins in the judgment of Nourse LJ in *John D Wood & Co* v *Dantata*,[2] where it was said that what the agent must achieve is an introduction, not to the property, but to the transaction. In effect, therefore, the word 'purchaser' means, not simply 'a person who at some time in the future becomes a purchaser' but rather 'a person who becomes a purchaser as a result of our introduction'. As we shall see,[3] there are some difficulties with this

---

1   [2008] 2 EGLR 23.
2   [1987] 2 EGLR 23.
3   P 124.

approach; however, it was relied upon by Lord Neuberger in *Foxtons* v *Pelkey Bicknell* to justify denying a firm of agents the commission which they claimed.

The principle in practice

The general rule stated above has to be applied by the courts to the infinitely variable facts of actual cases. Since, as already mentioned, it is a question of fact whether or not the agent is an effective cause, decisions on this point are not of binding authority for the future. Nevertheless, a consideration of some of these cases is instructive for two reasons: first, because certain general principles emerge from even the most empirical of inquiries, and second, as a guide to the types of case in which this point is likely to arise and the circumstances in which the agent may lose commission which at first sight appears to have been earned.

For an agent's act to rank as an effective cause of the event which earns commission, it must be carried out in pursuance of a commission agreement. It follows that there can be no successful claim in respect of an introduction made by the agent before the commission agreement is entered into.[1] Nor can an agent rely on an introduction made by a sub-agent, where the client has given the agent no authority to sub-instruct.[2] However, an introduction may be treated as the effective cause of a sale whether or not the client is aware of it, as for example where a prospective purchaser deceives the vendor by pretending to have come privately.[3] Furthermore, if a potentially effective introduction is made by an agent acting under a commission agreement, it will not be rendered ineffective by the vendor's decision to take over the negotiations in person[4] or through another agent.[5] Nor can the vendor avoid liability to pay commission by terminating the agent's instructions before concluding the deal,[6] although a sole agent

---

1   See p 139.
2   *McCann & Co* v *Pow* [1975] 1 All ER 129; *Robert Bruce & Partners* v *Winyard Developments* [1987] 1 EGLR 20.
3   See *Warman* v *Newmans Ltd* (1901) 17 TLR 509.
4   *Wilkinson* v *Martin* (1837) 8 C & P 1.
5   *Bow's Emporium Ltd* v *A R Brett & Co Ltd* (1927) 44 TLR 194, 197 *per* Lord Shaw of Dunfermline.
6   *Green* v *Bartlett* (1863) 14 CBNS 681; *Christie & Co* v *Jones* (1966) 198 EG 1093.

may not be entitled to claim on a sale to someone introduced within that period, where contracts are not exchanged until after the sole agency has expired.[1]

The question whether or not an introduction ranks as the 'effective cause' of a transaction is one which is answered objectively. Thus in *Burchell* v *Gowrie & Blockhouse Collieries Ltd*,[2] an agent was held entitled to commission on a sale to a person introduced, notwithstanding that the agent had advised the client against selling to this person. However, where an agent's reason for advising the client not to accept an offer from X was a secret agreement between the agent and Y, a rival applicant, one of the court's reasons for rejecting the agent's claim to be paid commission on a sale to X was that this sale had not been brought about by the agent.[3]

A situation which frequently gives rise to an effective cause dispute is that in which an agent 'introduces' a person who has already been in negotiation with the vendor. In order to succeed in such a case, the agent must show that the previous negotiations had come to nothing and that it was only because of the agent's intervention that they were revived.[4] An agent who is unable to establish this is unlikely to convince a court that things done subsequently, such as persuading the client to reduce the asking price, is decisive enough to earn commission.[5]

In many cases of this kind, a client is faced with claims from two or more agents. The strongly held (and strongly expressed) view of Lord Denning was that there could only ever be one commission per transaction, and that this would be earned by whichever of the agents found the ultimate purchaser.[6] However, there seems no reason in principle why more than one agent should not be entitled to commission in respect of the same transaction and, as a result, the

---

1   So held in *Fairvale Ltd* v *Sabharwal* [1992] 2 EGLR 27 (p 133), where the commission clause spoke of sales 'effected'. See also the doubts expressed by Simon Brown J in *Brodie Marshall & Co* v *Carpenter* [1985] 1 EGLR 24. However, the definition of 'sole agency' provided by the Estate Agents (Provision of Information) Regulations 1991 would entitle the agent to commission in such a case: see p 130.

2   [1910] AC 614.

3   *Henry Smith & Son* v *Muskett* [1978] 1 EGLR 13.

4   *Thompson, Rippon & Co* v *Thomas* (1895) 11 TLR 304; *Hampton & Sons* v *Trade & General Securities Ltd* [1979] 1 EGLR 39.

5   *Lewis* v *Calder* (1957) 170 EG 5.

6   See, for example, *AA Dickson & Co* v *O'Leary* [1980] 1 EGLR 25.

courts will not allow a client to interplead, that is, pay one amount of commission into court and leave the competing agents to fight over it.[1] How such double liability might come about was discussed by Drake J in *Lordsgate Properties Ltd* v *Balcombe*:[2]

> It appears to me that there is no good reason in law why a vendor may not be liable to two agents in respect of the same transaction, provided either (a) both parties were instrumental in causing the sale; or (b) the different contracts entitle each of the agents to commission for different reasons.

Taking the second of these possibilities first, it is clear that an agent instructed on 'willing to purchase' terms may succeed in earning commission for having made a suitable introduction, even where the property is ultimately sold to an applicant introduced by another agent. Again, it was held in the *Lordsgate* case itself that, where one agent was instructed to 'introduce an applicant who purchases' the property, and another to be 'instrumental in negotiating a sale', their claims were not mutually exclusive and indeed that both agents were entitled.

The first of the possibilities mentioned by Drake J is, however, rather more problematical. While in theory a court might decide that a purchaser has been introduced by the combined efforts of two or more competing agents, and that these agents must therefore share the commission, this approach does not appear to have been adopted in any reported case. More typical is the blunt statement of Lopes LJ in *Barnett* v *Brown*[3] that: 'The question to be decided was: Whose introduction had brought about the purchase?', although the modern judicial tendency is to pay lip service to the possibility of a double claim, while denying it on the facts of a particular case.[4]

When called upon to select an 'effective cause' from two or more competing claims, the courts profess to be completely open-minded. As Harman LJ put it in *Bartlett* v *Cole*:[5]

---

1   *Greatorex* v *Schackle* [1895] 2 QB 249; *Moss Kaye & Roy Frank & Co* v *Jones* [1979] CLY 37.
2   [1985] 1 EGLR 20.
3   *Barnett* v *Brown* (1890) 6 TLR 463.
4   See, for example, *Lordsgate Properties Ltd* v *Balcombe* [1985] 1 EGLR 20; *John D Wood & Co* v *Dantata* [1985] 2 EGLR 44; [1987] 2 EGLR 23.
5   (1963) 188 EG 397.

I do not think you can answer the question by saying: 'Who was first on the list?' I think you can answer the question by saying: 'Whose introduction was the effective introduction?'

Notwithstanding such expressions of opinion, however, it cannot be denied that the decisions have a tendency to favour the first introduction,[1] unless there is some very clear reason for this to be displaced.[2] An example of such a clear reason may be seen in *Egan Lawson Ltd* v *Standard Life Assurance Co*,[3] where the claimants told the defendants about an investment property and suggested a 'pre-emptive offer' of around £13m. The defendants were not interested at this price but, on later hearing from other agents that the vendors might accept a lower offer, they made one and ultimately purchased the property for £11.5m. It was held by the Court of Appeal that the effective cause of the sale was the information from the second agents, not the introduction by the claimants.

At first instance in *John D Wood & Co* v *Dantata*,[4] Forbes J said:

I do not consider that an agent who effects a second introduction to the property (if that is not a contradiction in terms) can succeed in demonstrating that such an introduction was the effective cause of the sale, unless he can show that the interest aroused in the purchase by the first introduction has evaporated by the time of the second.

Although this *dictum* was referred to without apparent disagreement when the *John D Wood* case reached the Court of Appeal, and was endorsed by Garland J in *Chesterfield & Co Ltd* v *Zahid*,[5] it was disapproved by the Court of Appeal in the case of *Chasen Ryder & Co* v *Hedges*.[6] According to Staughton LJ, the true position is rather more open:

---

1 See, in addition to *Bartlett* v *Cole* (1963) 188 EG 397, *Moss Kaye & Roy Frank & Co* v *Jones* [1979] CLY 37; *Glentree Estates Ltd* v *Gee* [1981] 2 EGLR 28; *Bentleys Estate Agents Ltd* v *Granix Ltd* [1989] 2 EGLR 21; *Your Move* v *Dunbar* [2001] CLY 112.

2 See *Taplin* v *Barrett* (1889) 6 TLR 30 (where the applicant subsequently bought the property at auction); *Robert Drummond* v *Mangles* [1981] 2 EGLR 31; *Chesterfield & Co Ltd* v *Zahid* [1989] 2 EGLR 24; *Cobbs Property Services Ltd* v *Liddell-Taylor* [1990] 1 EGLR 49.

3 [2001] 1 EGLR 27.

4 [1985] 2 EGLR 44; affirmed by the Court of Appeal [1987] 2 EGLR 23.

5 [1989] 2 EGLR 24.

6 [1993] 1 EGLR 47. This passage was approved and applied in *HRL Property Management Services Ltd* v *Chequers of Kensington Ltd* (27 October 2000).

The burden is on the claimant to show that his introduction in any case was the effective cause of the purchase. If, however, he shows that he was the first to introduce the purchaser, and that a purchase followed, then it may well be that the judge will infer that the claimant was the effective cause. It can therefore be said that the evidential burden in such a case passes to the defendant, whether the other agent or the vendor, to prove more facts which displace that inference. But even in such a case, I do not think that the further facts which the defendant then has to prove must be such as to show that interest aroused by the first introduction has evaporated, that is to say, entirely disappeared. It will be a matter for consideration in each case, how far the defendant has to go before he has displaced an inference which might arise from the mere fact of the introduction followed by the purchase.

Most of the 'two agent' cases cited above arose when applicants who had been shown a property by one agent lost interest, but were subsequently reintroduced to it by a second agent. However, an unusual example occurred in *Calloway & Co v Peters*,[1] where the claimants and another firm of estate agents had both erected 'For Sale' boards outside the defendant's house. A prospective purchaser, having made some enquiries through the claimants, told a member of her staff to contact 'the agents' and negotiate a purchase. The employee inadvertently telephoned the other firm, the sale was concluded through them, and it was held that the claimants were not entitled to commission.[2] Even more bizarre is the case of *Bloom v Yefet*[3] where, it appears, a Saudi Arabian princess was shown round a property for sale by two rival agents at the same time, each agent being blissfully unaware of the other's presence!

The 'effective cause' problem arises in a rather different form where there is a substantial delay between the original introduction and the ultimate transaction. Sometimes the break in causation is clear, as it was in the New Zealand case of *Lewis v Wong*.[4] The agents there showed a prospective purchaser several properties, one of which he duly bought. Not surprisingly, they were held not to be entitled to commission when, after negotiating directly with the vendor, he later bought one of the

---

1   [1954] 164 EG 353.
2   It is possible that a 'two board' case might today be decided according to which board was covered by deemed planning consent: see p 333.
3   [1982] 2 EGLR 30.
4   [1982] Recent Law 336. See also *Barnett v Brown* (1890) 6 TLR 463.

other properties. Less obvious are those cases in which a person introduced by the agent as a tenant of the client's property later purchases the freehold. In *Toulmin* v *Millar*[1] the agent's claim failed because, it was found, the authority given by the client was limited to finding a tenant. The question arose in a pure form in *Nightingale* v *Parsons*,[2] where the claimant was asked to find either a tenant or a purchaser for the defendant's house. A tenant was introduced who, after three years, purchased the freehold instead of exercising his option to renew the lease. It was held that the agent, who had played no part in the negotiations leading up to the sale, could not claim to be an 'effective cause' of it. In the county court case of *Poulter* v *Doggett*,[3] on the other hand, where the estate agent's instructions were similar, the lease itself gave the tenant an option to purchase the freehold. It was held that, when this option was exercised by an assignee of the original tenant, the estate agent was entitled to commission on the sale, less the amount which the client had already paid.

The question of commission on a subsequent transaction arose in rather different circumstances in *Tribe* v *Taylor*,[4] a case which emphasises the factual nature of this whole matter. The claimant there, who was commissioned to find capital for the defendant's business, introduced a person who lent £10,000, on which the claimant duly received commission. A few months later this person entered into partnership with the defendant and injected a further £4,000 into the business. It was held that this further loan resulted, not from the claimant's introduction, but from the partnership negotiations and that, accordingly, no commission was payable.

Exclusion of the requirement

As already noted, the effective case requirement may be excluded by the provisions of a particular contract or the facts of a particular case. It appears from the case law that this is most likely to occur in a case concerning commercial property (where issues of consumer protection do not arise). In *Brian Cooper & Co* v *Fairview Estates (Investments) Ltd*,[5] for example, the claimant agents were instructed to market an office

---

1   (1887) 3 TLR 836.
2   [1914] 2 KB 621. See also *Millar, Son & Co* v *Radford* (1903) 19 TLR 575.
3   (1964) 115 LJ 76.
4   (1876) 1 CPD 505; see also *Favermead Ltd* v *FPD Savills Ltd* [2005] EWHC 626 (Ch).
5   [1987] 1 EGLR 18.

development on terms that commission would be payable 'should you introduce a tenant ... with whom we have not been in previous communication and who subsequently completes a lease'. The claimants introduced a company which ultimately took a lease of the property, but only after its negotiations with the clients had ceased and been revived by another firm of agents. The claimants were clearly not the effective cause of the lease; but the Court of Appeal held that, on the wording of their commission terms, they were not required to be.[1]

Another area in which agents have enjoyed some success in excluding the effective cause requirement is that of seeking property on behalf of a prospective purchaser or tenant. An example is *County Homesearch Co (Thames & Chilterns) Ltd* v *Cowham*,[2] where the claimant agents were retained by the defendant to search for a suitable residential property. The claimants' terms of business made provision for commission on various events ('deemed introductions') which were not really introductions at all, something which the Court of Appeal regarded as inconsistent with an implied requirement of effective cause. This was enough to support the agents' claim, but Longmore LJ also pointed out that one of the arguments normally used to support the effective cause rule, that it protected the client from becoming liable for two commissions, was not relevant when acting for a purchaser, since such purchasers would normally engage only one agent.

As to other attempts to outflank the effective cause principle, it has been held that a term that commission is payable 'should a sale be effected' still requires that sale to be effected by the agent;[3] not so, however, where commission is tied to a sale 'whether arranged by the auctioneers or not'.[4] Even less explicit wording may play its part. In *Bayley* v *Chadwick*,[5] for example, where the claimant was employed to sell a ship by auction, his client agreed to pay commission if a private sale came about 'in consequence of your mention or publication for

---

1   *Raja* v *Rollerby Ltd* [1997] EGCS 79, a vendor was held liable on an agreement to pay in the event of a sale to 'your client', even though it transpired that the 'client' in question had already been shown the property by another agent.

2   [2008] 1 EGLR 24. In *MSM Consulting Ltd* v *United Republic of Tanzania* [2009] EWHC 121 (QB), by contrast, Clarke J was prepared to imply an 'effective cause' term into an agency agreement by which the claimants were to find suitable property for occupation as an embassy.

3   *Sadler* v *Whittaker* (1953) 162 EG 404.

4   *Bernard Thorpe & Partners* v *Snook* [1983] 1 EGLR 37; *Barnard Marcus & Co* v *Ashraf* [1988] 1 EGLR 7.

5   (1878) 39 LT 429.

auction purposes'. Information obtained from the claimant was passed on to a person who bought the ship direct from the client. The House of Lords held that, on the wording of this clause, the claimant was entitled to commission.[1]

In *Chris Hart (Business Sales) Ltd* v *Mitchell*,[2] the claimants' standard terms of business provided that, even after termination of their sole agency, they would be entitled to commission on any sale to which they had 'contributed in any way'. It was also provided that the client must forward to the agents any offer to purchase, failing which they would again be entitled to their full commission. When a client, having terminated the claimants' sole agency, marketed his hotel himself and found a purchaser, the agents' claim for commission failed. The mere fact that the defendant had based his sale particulars on those drafted by the agents did not mean that they had 'contributed' to the sale. Nor could they claim for failure to submit the offer to them; this was effectively a claim for damages unrelated to any loss suffered, which meant that the term was an unenforceable penalty clause.

Since 1991, any estate agent using the terms 'sole agency' or 'sole selling rights' in their terms of business has been under a statutory obligation to give the client a written explanation of these terms, using a form of words set out in the Estate Agents (Provision of Information) Regulations 1991. In the 4th edition of this book the view was expressed that this statutory wording excluded the requirement of effective cause, and this view was accepted by Nelson J in *Fleurets Ltd v Dashwood*.[3] However, in *Foxtons Ltd* v *Pelkey Bicknell*,[4] it was held by Lord Neuberger that, far from excluding the effective cause requirement, the regulations specifically included it; this was because the statutory phrase 'a purchaser introduced by us' meant 'a purchaser introduced by us to the transaction', which in turn meant someone who only became a purchaser as a result of our introduction.[5]

---

1. *Contrast Coles* v *Enoch* [1939] 3 All ER 327, where the ultimate purchaser merely overheard the agent discussing the property with someone else.
2. 1996 SCLR 68.
3. [2007] 2 EGLR 7.
4. [2008] 2 EGLR 23.
5. This interpretation of the words 'introduce a purchaser' was based on the views expressed by Nourse LJ in *John D Wood & Co* v *Dantata* [1987] 2 EGLR 23. However, it does not appear possible to apply an equivalent interpretation to an agreement under which an agent is to seek suitable property for a prospective purchaser: *MSM Consulting Ltd* v *United Republic of Tanzania* [2009] EWHC 121 (QB).

It is submitted that, while persuasive, this approach creates serious problems of analysis. Even if one agrees with Lord Neuberger's preferred interpretation of the phrase 'introduce a purchaser',[1] the problem is that previous courts have treated the effective case requirement as based, not on the interpretation of the express words used in the agency agreement, but on a term properly to be implied into that agreement. If his lordship is correct, it means that, at least in cases where similar words were used, the earlier courts' consideration of implied terms was completely irrelevant. More serious, the ruling appears in direct conflict with the actual decision of the Court of Appeal in *Brian Cooper & Co* v *Fairview Estates (Investments) Ltd*,[2] where similar wording was used. It was there held that, since no 'effective cause' term could be implied, the claimant agents must inevitably be entitled to their commission; the possibility that the wording itself might import an effective cause requirement was never even considered by the Court of Appeal.

Negotiation

The commission clauses which we have so far considered in this chapter, although depending upon a variety of events, nevertheless have one thing in common. Implicit in them all is the assumption that it must be the estate agent who actually makes the introduction of the applicant to the client. We must now turn our attention to a rather different function which an estate agent is not infrequently employed to carry out, namely, to *negotiate* on behalf of a client in circumstances where the parties have already been introduced, either privately or by another agent.[3] Many of the commission terms in common use (for example the former RICS scales) make express provision for this situation, and the precise requirements for a claim on such terms to succeed have been considered by the Court of Appeal on two occasions. In *F P Rolfe & Co* v *George*[4] the claimants, who had the defendant's

---

1   It may be noted that Lord Neuberger himself acknowledged that many might find the other interpretation (ie that adopted by legal writers, including the present one) a more natural one.

2   [1987] 1 EGLR 18; see p 121.

3   This does not constitute 'estate agency work' for the purposes of the Estate Agents Act 1979: see p 272.

4   (1969) 210 EG 455. See also *Anscombe & Ringland Ltd* v *Watson* [1991] 2 EGLR 28.

grocer's shop on their books, were called on by the defendant to negotiate on his behalf with a person who had been introduced by someone else. The claimants explained both the business and the accounts to this applicant and persuaded him to go through with the deal, but their claim to commission was rejected by the county court judge on the ground that they had not introduced the purchaser. When the claimants appealed (successfully) to the Court of Appeal, Lord Denning MR had this to say about the phrase 'negotiating a sale by private contract' which formed part of the claimants' terms of business:

> Those words mean, as I understand it, that the agent must be, by reason of his negotiation, the efficient cause of the sale; or ... he must have produced the sale by his negotiation. There was clear evidence that the agent did this.

It may at first sight seem odd that an agent who by definition has not introduced the parties must nevertheless show that he or she is the 'effective cause' of the sale, but this requirement is in fact a perfectly reasonable one. After all, what the agent is claiming is not a fee for professional services, but commission based on the value of the property; and this, as discussed earlier,[1] depends upon the achievement of a result.

The second requirement for success under a clause of this type is more controversial. In *Hoddell* v *Smith*,[2] an agent's claim for commission was rejected by the Court of Appeal for reasons which the following extracts from the judgments make clear. First, in the opinion of Geoffrey Lane LJ:

> 'Negotiation' must, to my mind, mean conferring with the prospective purchaser, or his representatives, with a view to agreeing a sale, and cannot include the simple giving of advice to the agent's own principal, the seller.

Sir John Pennycuick was even more explicit:

> I do not think the word 'negotiate' is apt to cover (i) the giving of advice by an agent to his principal where the principal is conducting personally or through another agent the negotiations with the other party concerned; (ii) preparing plans and the like in connection with such

---

1   P 82.
2   [1976] 2 EGLR 38.

negotiations being carried on by the principal or another agent; (iii acting as a mere channel of communication between the principal and his prospective purchaser; or (iv) a mere willingness to enter into negotiations if so required by the principal.

It is submitted that to require the agent to 'negotiate' in this formal sense is both illogical and unjust, for an agent may be wholly effective in bringing about a sale while standing behind, rather than in front of, the client. Nevertheless, the principle has been applied in two subsequent cases,[1] where agents acting on behalf of companies wishing to rent business premises were to be remunerated for 'seeking and negotiating a tenancy or lease'. In each case the agents found suitable premises and a lease was duly signed, but the absence of any positive 'negotiations' was held to mean that there could be no claim for commission as such. True, the court in each case was able to award the agents a 'reasonable sum' for their work but, as we shall see,[2] these sums fell far short of what would have been payable under the agents' scales.

Sole agency

The common law position

A 'sole agency' agreement is one where, in return (it appears) for an express or implied promise of services,[3] an estate agent is granted the exclusive right to introduce a purchaser for the vendor's property. The extent of the rights given to an estate agent by such an agreement, which of course restricts the vendor's freedom of contract, has been considered by the courts on a number of occasions. A straightforward example is provided by the case of *Hampton & Sons Ltd* v *George*,[5] in which the claimants were appointed sole agents for the sale of the defendant's leasehold hotel. Another firm had had the property on their books for some time but their authority was determined on the appointment of the claimants. Nevertheless the other firm introduced

---

1   *Reiff Diner & Co* v *Catalytic International Inc* [1978] 1 EGLR 16; *Sinclair Goldsmith* v *Minero Peru Comercial* [1978] 2 EGLR 28.

2   P 136.

3   The estate agent's obligations are considered at pp 46–49.

4   The mutual obligations of 'joint sole agents' were considered by the Court of Appeal in *Hampton & Sons* v *Garrard Smith (Estate Agents) Ltd* [1985] 1 EGLR 23.

5   [1939] 3 All ER 627.

an applicant whose offer was accepted by the defendant. The claimants sued for and were awarded damages for the defendant's breach of the sole agency agreement.

In *Bentall, Horsley & Baldry* v *Vicary*[1] the claimants were appointed sole agents for a period of six months for the sale of the defendant's house. During this period the defendant sold the house privately, not through any agent. It was held that, in the absence of an express prohibition, the defendant was quite entitled to do this, and so the agents could not recover either commission or damages. McCardie J distinguished the earlier case of *Chamberlain & Willows* v *Rose*,[2] which concerned the sale of a cinema. The claimants in that case were appointed to sell the property by auction or otherwise under an agreement which stated 'The property to be left solely in your hands for sale from this date until the auction and for a further period of three months.' It was held that the defendant was liable in damages for breach of this agreement when, during the period in question, he sold the cinema privately.

It is clear, then, that an agent who is given 'sole selling rights' is in a better position than one who is appointed 'sole agent', since the former is protected, not only against sales through other agents, but also against private sales by the vendor. The category into which a particular agent falls is to be determined by interpreting his instructions, and it is suggested that the phrase 'sole selling agent'[3] should in principle be interpreted as conferring merely a sole agency and not sole selling rights.[4] Where either category is concerned, however, it is submitted that, notwithstanding certain remarks of Ewbank J in *Glentree Estates Ltd v Gee*,[5] a mere withdrawal of the property by the vendor is not a breach; it is only a sale which may be wrongful.

At common law, a claim for breach of a 'sole agency' or 'sole right to sell' is a claim for damages, not commission. As a result, such questions as whether the agent's instructions have been fulfilled, or whether the agent is the effective cause of the sale, do not arise. Moreover, since the agent's complaint is of being deprived of the opportunity to sell the property, the damages awarded should reflect

---

1   [1931] 1 KB 253.

2   (1924), unreported but summarised at [1931] 1 KB 261.

3   Or 'sole letting agent': see *Folioshield Ltd* v *Pleamere Ltd* [1990] 2 EGLR 1.

4   Although certain remarks in *Brodie Marshall & Co (Hotel Division) Ltd* v *Sharer* [1988] 1 EGLR 21 lend some support to the opposite view.

5   [1981] 2 EGLR 28, 29.

the value of that lost opportunity, and will thus not necessarily equal the full commission which would have been earned. In *Hampton & Sons Ltd* v *George*,[1] for example, the claimants' commission on the sale of the lease would have amounted to £104. However, due to the difficulties surrounding the sale of licensed premises, in that the landlords, a brewery company, had to give their consent, it was by no means certain that a suitable purchaser would have been found. Accordingly damages were assessed at £80.[2]

The case of *Morris Oddy Ltd* v *Hayles*[3] appears rather at odds with these principles of compensation. The defendant in that case, who was developing an estate, appointed the claimants as his selling agents under an agreement, the important terms of which were as follows: 'Upon a sale being effected by you, I hereby agree to pay you a commission at the rate of £75 for each bungalow sold ... You are to be sole agent in the sale of the said properties ... All inquiries received by me will be referred to you, and I shall not have any rights of direct negotiations with purchasers.' The agreement was expressed to be irrevocable for a period and thereafter terminable by notice of a specified length. A number of bungalows were sold in clear contravention of this agreement and the claimants accordingly sued. It seems reasonably clear from the judgment of Thesiger J that he awarded the claimants their agreed commission on all sales up to the date on which the agreement was validly terminated. If this is so, the award was surely incorrect, since by no stretch of the imagination could these sales be said to have been 'effected' by the claimants, as the agreement called for. However, the sum awarded could also be reached by another route, which would avoid this logical inconsistency: since the agreement stipulated that all inquiries should be referred to the claimants, it was a near certainty that those which led to a sale would have done so anyway; thus damages based on the full commission could validly be awarded.[4]

What is effectively a sole agency or sole right to sell may on occasion be created without using these actual terms. A device

---

1   [1939] 3 All ER 627.
2   In *Newton* v *Eriksson* (1951) 157 EG 414, where commission would have been £115, damages of £75 were awarded. In *HM Rendall* v *Lammas* [1967] CLY 36 the 'discount' was a mere £10.
3   (1971) 219 EG 831.
4   See *Gross Fine & Krieger Chalfen* v *Gaynor* [1975] 1 EGLR 38, where a buoyant market justified an award of damages equivalent to the full commission.

frequently encountered is a stipulation that commission shall be payable upon any sale which takes place during the currency of the sole agency agreement. Less widely used, but successful before the Court of Appeal in *Property Choice Ltd* v *Fronda*,[1] was an express undertaking by the client not to 'consent to sell the property to anyone not introduced by [the agents]'. This, it was held, entitled those agents to damages on any such sale, whether negotiated through another agent or privately by the client.

Such variations on the sole agency theme require careful drafting. It has for example been held that, where an agent was to receive commission 'should a sale be effected', this meant 'effected by the agent'; there was accordingly no right to claim when the vendor sold the property privately.[2] However, this problem can be overcome by making commission payable 'if a sale of the property, whether arranged by the [agents] or not, is effected' within a stated time.[3] Such wording is today commonly found in auctioneers' commission clauses, and is further discussed in that context.[4]

The 1991 Regulations

Everything which has so far been said about sole agency and its variants must henceforth be read in the light of the Estate Agents (Provision of Information) Regulations 1991, which were made under section 18 of the Estate Agents Act 1979. Those regulations require any estate agent using the term 'sole agency', 'sole selling rights' or an equivalent, to explain to the client in writing what the term means, using the forms of explanation contained in the regulations themselves. Significantly, the definitions given appear to put the agent in a more favourable position than would be the case under the general law.

According to the 1991 Regulations, 'sole agency' means that remuneration will be payable in the following situation:

> if at any time unconditional contracts for the sale of the property[5] are exchanged with a purchaser introduced by us during the period of our

---

1   [1991] 2 EGLR 249.
2   *Sadler* v *Whittaker* (1953) 162 EG 404.
3   *Bernard Thorpe & Partners* v *Snook* [1983] 1 EGLR 37; *Barnard Marcus & Co* v *Ashraf* [1988] 1 EGLR 7. See, however, *Fairvale Ltd* v *Sabharwal* [1992] 2 EGLR 27.
4   See p 133.
5   As to what is meant by 'the property' for this purpose, see *Dowling Kerr Ltd* v *Scott* [1996] EGCS 177: p 111.

sole agency or with whom we had negotiations about the property during that period; or with a purchaser introduced by another agent during that period.

A number of points may be made about this clause:

- Commission is explicitly tied to the exchange of contracts thus ruling out any implied requirement for the person introduced to remain willing and able to complete the sale.[1]
- The agent will be entitled to commission whenever the sale occurs, so long as the relevant introduction (whether made by the agent or by another agent) takes place during the period of sole agency.[2]
- An 'introduction' for this purpose merely requires the agent to have brought the property to the purchaser's attention (for example by sending sale particulars); it is not necessary to show that the agent brought the parties into direct contact with each other.[3]
- In the 4th edition of this book the view was expressed that the wording of the 1991 Regulations was such that, wherever it was used in an agreement for sole agency or sole selling rights, it would remove the normal requirement that the agent be the 'effective cause' of the relevant transaction. This view[4] was specifically accepted by Nelson J in *Fleurets Ltd* v *Dashwood*.[5] However, it was specifically rejected by Lord Neuberger, sitting in the Court of Appeal, in *Foxtons Ltd* v *Pelkey Bicknell*.[6]
- The agent will be entitled to full commission, rather than an award of damages for breach of the sole agency, which would normally reflect only the lost chance of earning commission.[7]

---

1   See p 94.
2   Contrast *Fairvale Ltd* v *Sabharwal* [1992] 2 EGLR 27: p 133.
3   *Christie Owen & Davies plc* v *Ryelance* [2005] 18 EG 148 (CS); *Christie Owen & Davies plc* v *King* 1998 SCLR 786. See also *Your Move* v *Dunbar* [2001] CLY 112.
4   Shared by *Bowstead on Agency* (18th ed, 2006, at p 279).
5   [2007] 2 EGLR 7.
6   [2008] 2 EGLR 23. For discussion of this ruling, see p 124.
7   See, for example, *Hampton & Sons Ltd* v *George* [1939] 3 All ER 627; p 128), where the agents recovered as damages approximately three-quarters of what would have been earned as commission.

Sole selling rights

The definition of 'sole selling rights' contained in the 1991 Regulations requires a client to pay remuneration in either of two situations. The first of these is:

> if unconditional contracts for the sale of the property are exchanged in the period during which we have sole selling rights, even if the purchaser was not found by us but by another agent or by any other person, including yourself.

The second commission-earning situation is:

> if unconditional contracts for the sale of the property are exchanged after the expiry of the period during which we have sole selling rights but to a purchaser who was introduced to you during that period or with whom we had negotiations about the property during that period.

Notwithstanding its different layout, this definition seems for the most part equivalent in its effect to that of 'sole agency' (apart, of course, from the fact that it catches private introductions as well as those made by a rival estate agent). In consequence, all the comments made above about the earlier definition apply with equal force to this one. However, there is one significant difference, identified in *Michael Harwood t/a RSBS Group v Smith*.[1] The claimants there were given sole selling rights over a residential nursing home for a period of six months. During that period the client himself found a potential purchaser, but waited until two days after the claimants' agency had terminated before exchanging contracts. The Court of Appeal held that the phrase 'purchaser who was introduced to you during that period', in the second commission-earning situation described above, means 'introduced by the agents themselves'; thus the client had successfully avoided having to pay commission.

Auctions — direct sale by vendor

A version of the 'sole agency' scenario occurs where property which has been put into the hands of an auctioneer for sale is sold privately by the vendor, either before the date of the auction or after the auction

---

1   [1998] 1 EGLR 5.

has failed to produce a purchaser. A claim by an auctioneer for commission in such circumstances was successful in the old case of *Rainy* v *Vernon*,[1] in which an auctioneer was employed to sell certain ground rents. The client sold these ground rents personally before the date of the auction, whereupon the auctioneer brought an action for commission, calling as witnesses three other members of the profession, who said that the custom in these circumstances was for commission to be payable. In summing up to the jury, Lord Denman CJ said:[2] 'If, however, you think that the witnesses called for the claimant have, by their evidence, made out a case of usage so notorious that the defendant was aware of it, then the custom is engrafted on the contract, and the claimant is entitled to his commission'. The jury was clearly persuaded by these witnesses, for it decided in favour of the auctioneer.

The custom which was established to the satisfaction of the jury in *Rainy* v *Vernon* does not appear to have formed the basis for any subsequent successful claims by auctioneers. Indeed, the whole trend of decisions since that time has been against awarding commission to an auctioneer who is not in some way personally responsible for bringing about a sale. In *Green* v *Hall*[3] the auctioneer's instructions provided for the payment of commission if the client's estate were sold. After an abortive auction, the client sold the property privately through another firm of agents. It was held that, since 'sold' meant 'sold by the auctioneer himself',[4] no commission was payable; the auctioneer was entitled only to the fee which had been agreed as payable in the event of no sale resulting.

In *Williams* v *Tuckett*[5] where the defendant, having instructed the claimant to sell property by auction, withdrew it from the sale and sold it privately, the jury were again asked to find a custom that commission was payable in these circumstances. However, although *Rainy* v *Vernon*[6] was referred to, the jury found that no such custom existed, after a summing-up by Lawrence J which indicated that requiring the auctioneer to have brought about the sale was 'consistent with common law and common sense'.

---

1   (1840) 9 C & P 559.

2   At p 562.

3   (1848) 12 LT (OS) 151.

4   A similar construction has been placed upon estate agents' agreements: see *Sadler* v *Whittaker* (1953) 162 EG 404.

5   [1900] Times, March 9. See also *Watson* v *Newton* (1902) 96 EG 140.

6   (1840) 9 C & P 559.

In the light of such judicial attitudes, it is hardly surprising that auctioneers have sought to include in their standard terms of business a clause to safeguard their commission in the event of the client withdrawing property for private sale. The following clause, a version of which was adopted by the Royal Institution of Chartered Surveyors, is typical:[1]

> Commission will amount to [a specified percentage of the sale price] and will be payable if a sale of the property, whether arranged by the auctioneers or not, is effected after the acceptance of instructions and before the auction, or within [a specified period] after the auction.

In *Bernard Thorpe & Partners* v *Snook*,[2] where a vendor sold his farm privately shortly after it had failed to reach its reserve at auction, it was held by Nolan J that a clause of this kind was effective to secure the auctioneers' commission. This decision was approved by the Court of Appeal in *Barnard Marcus & Co* v *Ashraf*,[3] where the defendant's property was sold through another estate agent some three weeks before it was due to be put up for auction by the claimants. The defendant argued that this constituted a 'withdrawal' of his property, for which the claimants would have been entitled under a separate clause to a mere £200 plus expenses, but the Court of Appeal ruled that commission was payable on the full price realised.

Although it would seem sensible for auctioneers to adopt terms of business similar to those under discussion, it must be appreciated that these do not cover every eventuality. In *Fairvale Ltd* v *Sabharwal*,[4] for example, where the defendant's hotel failed to reach its reserve price and was bought in, the claimant auctioneers introduced their client to one of the unsuccessful bidders. This introduction, which was made immediately after the sale, led to a private sale of the property to that bidder. When the auctioneers sought commission on this sale, relying on a clause which required a sale to be effected within 28 days of the abortive auction, it was held that their claim must fail; the word 'effected' meant that, to qualify, the actual contract of sale must be entered into within the 28-day period.

---

1   See *Barnard Marcus & Co* v *Ashraf* [1988] 1 EGLR 7; *Fairvale Ltd* v *Sabharwal* [1992] 2 EGLR 27.
2   [1983] 1 EG 37.
3   [1988] 1 EGLR 7.
4   [1992] 2 EGLR 27.

A second limitation on such clauses, albeit one which is unlikely to cause frequent problems, was exposed by the case of *John Meacock & Co* v *Abrahams*.[1] The claimants there were instructed to sell property by the defendant, who was second mortgagee. Immediately before the auction, the mortgagor sold the property and paid off the mortgage, which of course left the defendant with nothing to sell. The county court judge held that the claimants were entitled to their commission under a clause of the kind under discussion, but this decision was reversed by the Court of Appeal, by a majority. The Court of Appeal's ruling was that this clause applied only to cases where the property was sold by the vendor or with his concurrence; it did not protect the auctioneer where (as here) the sale was effected by a third party over whose actions the vendor had no control.

Amount of commission

Once it is established that an estate agent has done what is necessary to earn commission, all that remains to be decided is the amount which is to be paid. In a limited number of cases, this is governed by law. In particular, where land is sold by order of the Chancery Division, Family Division, Court of Protection or a divorce county court, the specific authorisation of the court is needed if the estate agent's charges (which must be inclusive of everything except surveys) are to exceed either that agent's normal sole agency rate or 2.5% of the sale price.[2]

Where no statutory scale applies, the amount of commission payable depends upon the agreement between agent and client. This may stipulate for a fixed sum, or (most unusually) provide that the agent is to receive any excess over a certain amount which the property realises, or it may incorporate a scale of charges issued by a professional society. As to such scales, it should be noted that, as a result of the Restriction on Agreements (Estate Agents) Order 1970,[3] which was passed to give effect to recommendations contained in a report of the Monopolies Commission, it is unlawful for estate agents to accept restrictions in respect of their charges in connection with the sale or purchase of unfurnished dwellings. Thus, although members of

---

1   [1956] 3 All ER 660.
2   *Practice Direction* [1983] 1 All ER 160. However, this 'scale' does not apply to sales of investment, business or farm property, nor to property sold in lots.
3   SI 1970 No 1696.

a professional body may choose to adhere to a recommended scale of charges in respect of residential property, no sanctions may be taken against them if they prefer not to.

In certain circumstances an agent may be entitled to claim a *quantum meruit*, or 'reasonable sum', for services rendered. This may arise, for example, where an express commission agreement has become unenforceable for some reason. In principle, it will also occur where there has been no express agreement as to commission,[1] since a mere request for services from a professional person raises the implication that these will be paid for at a reasonable rate of remuneration.[2] In *Miller v Beal*[3] the defendant, an auctioneer, who had advanced £1,500 to a man named Elliott, was empowered to sell Elliott's property to recover his money. The claimant, another creditor, laid claim to part of the property, whereupon it was agreed that the sale should take place and the money should be held in trust for the parties jointly. No express provision was made for the defendant to receive commission on the sale, but the court held that a reasonable fee could be charged.

Where an estate agent is entitled to claim a reasonable sum, the appropriate amount will be assessed by a court. In accordance with the principles discussed earlier,[4] the sum awarded should logically be based on the idea of 'participation' rather than on that of 'time and trouble'.[5] This method of assessment receives support from a number of cases concerning commission agents in the field of commerce.[6] For estate agents, too, a 'reasonable sum' has frequently been geared to professional scales or the usual commission rate, at least where the evidence shows clearly that the parties intended remuneration to take the form of commission rather than a fee.[7] However, a certain amount

---

1   Such cases should be rare, in view of the statutory obligation to notify the client of all potential charges: see p 287.

2   *Manson* v *Baillie* (1855) 2 Macq 80. This implied term now has statutory force: Supply of Goods and Services Act 1982, s 15.

3   (1879) 27 WR 403.

4   P 82.

5   Estate agents themselves do not always recognise this and may seek to base a *quantum meruit* claim on an hourly rate: see, for example, *MSM Consulting Ltd* v *United Republic of Tanzania* [2009] EWHC 121 (QB).

6   *Way* v *Latilla* [1937] 3 All ER 759; *Campbell* v *National Trust Co Ltd* [1931] 1 WWR 465; 1 DLR 705.

7   See, for example, *Turner* v *Reeve* (1901) 17 TLR 592; *Lewis & Graves* v *Harper* [1979] 1 EGLR 43; *Michael Elliott & Partners Ltd* v *UK Land plc* [1991] 1 EGLR 39; *Benhams Ltd* v *Kythira Investments Ltd* [2004] EWHC 2973 (QB).

of confusion has been caused in this connection by judicial remarks suggesting that what an estate agent has done is not 'worth' the amount of commission claimed. In *Hampton & Sons* v *Trade & General Securities Ltd*,[1] Pain J (who was forced reluctantly to concede that the claimants had bound their client to pay in accordance with the RICS scales) said that what they had done, 'although undoubtedly valuable, would from most people's point of view hardly seem to be £27,500 worth of work'. So too, in *Lewis & Graves* v *Harper*,[2] where the claimants claimed commission of £276, the Court of Appeal thought that a 'reasonable sum' for their services would have been £60.

The remarks so far mentioned were all *obiter*, so that the views expressed did not actually affect the amount of commission awarded. In a number of other cases, however, estate agents who had undeniably achieved what their clients wanted were remunerated on a much lower basis than the relevant professional scale. In two such cases, agents had found suitable business premises for their clients but failed to qualify for commission in accordance with their terms of business, as they had not 'negotiated' on their clients' behalf.[3] Having held that the relevant clause, on its true interpretation, expressly provided for the agents to receive a reasonable sum in this situation, the judge in each of these cases was faced with the problem of assessing that sum. In *Reiff Diner & Co* v *Catalytic International Inc*,[4] the possibility was considered that the fee for merely 'seeking' (and finding) a tenancy should be 2.5% of one year's rent, which was the difference between the scale fees for 'seeking and negotiating' and simply 'negotiating'. While rejecting this interpretation, the deputy judge nevertheless assessed the claimants' reasonable sum at the very figure which such a calculation would have produced, describing this result as 'coincidental and irrelevant'! In *Sinclair Goldsmith* v *Minero Peru Comercial*,[5] O'Connor J declined even to refer to the scale in question; the figure which he declared to be a 'reasonable sum' was approximately one-third of what would have been earned on a scale.[6]

---

1   [1979] 1 EGLR 43. See also *Luxor (Eastbourne) Ltd* v *Cooper* [1941] AC 108, 125, *per* Lord Russell of Killowen.

2   [1979] 1 EGLR 43.

3   On this point see p 124.

4   [1978] 1 EGLR 16.

5   [1978] 2 EGLR 28.

6   In *Debenham Tewson & Chinnocks plc* v *Rimington* [1989] 2 EGLR 26, failure to provide for a particular contingency meant that the claimants received £15,000 for their services instead of £487,500 as commission!

A similar lack of judicial generosity towards estate agents can be seen in cases where no written agreement of any kind has been entered into between agent and client. In *Chaskill Ltd* v *Marina Developments Ltd*,[1] which concerned the sale of a marina, the judge explicitly costed out the time spent by agents in effecting an introduction, arriving at a figure of £2325, which fell a long way short of the £15,000 which the agents regarded as a reasonable commission. And in *Withey Robinson* v *Edwards*,[2] although no such costing exercise was carried out, a firm of surveyors who assisted a client in the purchase of a night club found that, because they had not stipulated for payment in accordance with RICS scales, their bill was reduced by approximately one-third.

It is suggested that the approach adopted in these cases is erroneous. If the view of the courts is that an estate agent is not entitled to be paid unless the property is sold, it should logically follow that a 'reasonable sum' in respect of a successful transaction must be considerably higher than would be justified by reference to 'time and trouble'.[3] Unless and until such arguments meet with judicial approval, however, estate agents should ensure that their terms of business are watertight, so that they are not driven to rely on claims for a 'reasonable sum'.

In deciding what is a 'reasonable sum' in the auction context, courts have sometimes had regard to what is customary in the profession, established either by the evidence of other practitioners or, perhaps, by reference to the scales of charges issued by the professional societies. In *Re Page*,[4] for example, a solicitor employed an auctioneer to sell a client's house, without making any express agreement as to the amount of commission payable. The client's executor challenged the auctioneer's charges, but these were allowed upon the evidence of several other auctioneers which showed that, if anything, the fees were below the average. However, the court's discretion in this matter is unfettered. In *Re Wolfe, Heller* v *Wolfe*,[5] where land was sold by order of the court for £192,500, the relevant scale did not extend beyond £25,000. The judge's attention was drawn to the scale for sales in bankruptcy; nevertheless, he awarded the auctioneers approximately

---

1   [1988] 2 EGLR 241.
2   [1986] 1 EGLR 32.
3   See, further, Murdoch: 'Professional Fees – How Much is Reasonable?' [1981] Conv 424.
4   (1863) 32 Beav 487.
5   [1952] 2 All ER 595.

80 per cent of what they would have earned on that scale, describing this as a 'proper figure'.

Reference to what is usual in the profession does not always work in favour of the auctioneer at the expense of the client. In *Newman* v *Richardson*,[1] for example, the claimant claimed commission upon the sale of a dry dock. The defendant was willing to pay the normal scale fee for the sale of property, but the claimant claimed to be entitled to approximately five times this amount, on the ground that the property was of a special nature and difficult to sell. The jury had no hesitation in finding that the scale represented a perfectly adequate remuneration for the claimant's services.

The practice of assessing what it is 'reasonable' to pay by reference to the prescribed professional fees seems quite acceptable in cases where an auctioneer has earned commission by successfully carrying out a sale. Where, however, no sale results and the auctioneer is, by agreement, to receive something for the abortive work involved,[2] it is suggested that a substantially smaller sum would be appropriate. This is because scale fees are not related to the amount of work required to earn them; they are higher than this, in order to make up for the large number of sales on which effort is expended but no commission is earned.

This aspect of professional scales has effectively been recognised by the courts in cases involving the breach of a 'sole agency' agreement, in that the damages awarded to the agent have usually been rather less than the commission which has been lost.[3] However, in exceptional circumstances it may be justifiable to use professional scales of charges as a basis for reasonable remuneration for services which are of no benefit to the client. In *Frank Swain (a firm)* v *Whitfield Corporation Ltd*,[4] for example, the claimant's authority to sell certain property was withdrawn some 10 minutes before the sale was due to commence, by which time there were 50 bidders present. The evidence showed that a sale would almost certainly have resulted; accordingly Phillimore J,[5] whose decision was unanimously affirmed by the Court of Appeal, awarded the claimants the scale commission on the reserve prices, which was felt to constitute a reasonable sum for their services.

---

1   (1885) 1 TLR 348.
2   Whether there would be any entitlement in the absence of such agreement is doubtful: see p 144.
3   See pp 127–128.
4   (1962) 183 EG 479.
5   (1962) 182 EG 11.

Formalities

In order to make a successful claim for commission, an estate agent must be able to show that the client agreed to pay on the event which has occurred. Many commission disputes turn on the issue of whether or not the client ever actually instructed the agent to act on his or her behalf in finding a purchaser (or a property).[1] This is a question of pure fact, on which some agents win[2] and more lose.[3] If there is any lesson to be learned from the cases, it is probably that claims based on an alleged agreement which is entirely oral are less likely to succeed.

In order to recover commission, the agent must show not only that the client agreed to pay commission, but also that this agreement was obtained before the commission-earning event took place; to hold otherwise would mean allowing the agent to rely on past consideration to enforce the client's promise.[4] As a result, where an 'agent' sends unsolicited information about a property to someone who the agent thinks may be interested in acquiring it, the recipient of that information is perfectly entitled to use the information without incurring any liability to pay commission. This harsh truth for non-retained agents who send particulars of a property as 'cold mail' was spelled out by Judge Cowell in *Lady Manor Ltd* v *Fat Cat Café Bars Ltd*:[5]

> If I (or an estate agent) in the absence of any agreement, tell another of a property which may suit that other I am not entitled at the same time or thereafter to require that other to pay any fee for the information ... The basic principle of the law ... is that if I confer an unrequested benefit upon another, whether it be by the doing of works or the provision of information or in some other way, I am not entitled to require payment from him, the recipient of the benefit ... The recipient may use the works done or information provided however he likes and without obligation.

---

1   A variation on this occurred in *Aylesford & Co (Estate Agents) Ltd* v *Al-Habtoor* [2003] EWHC 2451 (Admin), where the defendant failed in his attempts to convince a judge that he had terminated the claimants' sole agency.

2   See, for example, *Benhams Ltd* v *Kythira Investments Ltd* [2004] EWHC 2973 (QB); *Day Morris Associates* v *Voyce* [2003] 12 EG 129 (CS).

3   See, for example, *Farrant* v *Unisys Europe-Africa Ltd* [1993] 2 EGLR 47; *George and Morris t/a Temple Trees* v *Bellway Homes Ltd* [2002] PLSCS 195; *Ian Green Residential Ltd* v *Asfari* [2007] EWHC 1491 (QB).

4   *Samuel & Co* v *Sanders Bros* (1886) 3 TLR 145, *per* Lord Esher. See also *Mote* v *Gould* (1935) 152 LT 347, 348, *per* du Parcq J; *Robert Bruce & Partners* v *Winyard Developments* [1987] 1 EGLR 20.

5   [2001] 2 EGLR 88.

It is therefore important to see what steps an estate agent can and should take to ensure that commission terms, whatever they may be, are binding upon the client.

Commission agreements

The best method of binding a client is undoubtedly to obtain his or her signature on a document in which all the relevant terms are set out; if this is done, then the general principles of contract law suggest that the client is bound, whether or not the document has been read.[1] However, some doubt has been cast upon the effectiveness of this procedure by Lord Denning MR, whose regard for the 'common understanding of mankind' (that commission comes out of the purchase price on completion of the sale) has led him to argue that an estate agent who seeks to depart from this understanding must ensure that the client has actual notice of the proposed alteration. The learned judge relied upon this principle as one ground for his decision in the case of *Jaques* v *Lloyd D George & Partners*,[2] which concerned one of the widest commission agreements ever to come before the courts. In that case the vendor signed a form which contained the following clause: 'Should you be instrumental in introducing a person willing to sign a document capable of becoming a contract to purchase at a price, which at any stage of the negotiations has been agreed by me, I agree to pay you a commission of £250 ...' The estate agents did not explain to the vendor the meaning of this clause and, indeed, their representative had already told him: 'If we find a suitable purchaser and the sale goes through, you will pay us £250.' The sale did not go through, the landlord having refused his consent to the assignment, and the Court of Appeal rejected the estate agents' claim to have earned their commission by virtue of the clause quoted above.[3] It was held that the vendor was not fixed with notice of what he had signed, as the effect of the document had been misrepresented to him.[4] Lord Denning MR, however, went further,

---

1   *L'Estrange* v *Graucob* [1934] 2 KB 394. Unless of course the client's letter is made conditional, for example by using the phrase 'subject to contract': see *Ronald Preston & Partners* v *Markheath Securities plc* [1988] 2 EGLR 23.

2   [1968] 2 All ER 187.

3   For the views of the Court of Appeal on the effect of this clause, see p 106.

4   Following *Curtis* v *Chemical Cleaning & Dyeing Co* [1951] 1 KB 805. The position would be the same if the client had been misled by an estate agent's advertisements: *Homebuyers Estates (Fylde) Ltd* v *Laycock* (1981, unreported).

stating that an agent who seeks to depart from the usual understanding of men must explain this to the client:

> In the absence of such explanation, a client is entitled to assume that the form contains nothing unreasonable or oppressive. If he does not read it and the form is found afterwards to contain a term which is wholly unreasonable and totally uncertain, as this is, then the estate agent cannot enforce it against the innocent vendor.[1]

If correct, this assertion would impose upon an estate agent a duty more onerous than that imposed upon other categories of agent, namely, that of ensuring that a client is given specific notice of the agent's terms of employment. This would be a departure from the normal position under the law of contract and, it is submitted, such an innovation cannot be supported. Indeed, it runs counter to an earlier decision of the Court of Appeal concerning estate agents which, unfortunately, was not cited to the court in *Jaques* v *Lloyd D George & Partners*. The case in question was that of *Pilkington* v *Flack*[2] in which a client, who was busy serving customers in her shop, was induced by an estate agent to sign a printed commission agreement. This document was described as 'scandalous' by the Court of Appeal, which none the less held that the client, having signed this document, was bound by what it contained.[3]

Confirming letters

Notwithstanding the benefits to be derived by obtaining written instructions signed by the client, many practising estate agents (especially in the residential field) have traditionally preferred to take instructions orally and then to 'confirm' these to the client in writing.[4] In so far as such a letter really does confirm points which have been specifically agreed between the parties, it is unlikely to cause any harm,[5] although whether a letter sent by an agent is of much positive value in

---

1   [1968] 2 All ER 187, 190.

2   (1948) 152 EG 366: see p 64.

3   A 'scandalous' agreement might today be struck down under the Unfair Terms in Consumer Contracts Regulations 1999; see p 148.

4   Whether this practice satisfies the Estate Agents Act 1979, s 18 is considered at p 293.

5   However, in *London Commercial & Land Ltd* v *Beazer Lands Ltd* [1990] 1 EGLR 54 the conditional wording of a letter ('we would look to you for payment') was held to negate any obligation to pay!

proving what the agent claims was agreed must be open to doubt. However, very considerable difficulties may arise where the letter is not an accurate record of the earlier meeting, either because it contradicts what was actually said or because it introduces fresh material. The problem lies in producing evidence of the client's agreement to the relevant terms, given the basic rule of contract law that silence is not consent.[1] Further, it remains equally intractable whether the relationship be classified as a unilateral or a bilateral contract.[2] After all, if the contract is bilateral, it is formed when instructions are first taken, so that a variation introduced by one party alone is not binding. If, on the other hand, the relationship consists of a unilateral contract, the offer is made by the client and the 'confirming letter' may be seen as a counter-offer from the agent, binding only when positively accepted.

The theoretical difficulties, then, are formidable, but the case law undoubtedly suggests that a confirming letter may none the less be of some effect. In the absence of express agreement to the contrary, it is established that the basis on which an estate agent will be assumed to operate is that of 'finding a purchaser' in return for a 'reasonable sum'. Both these elements have been held by the Court of Appeal to be capable of alteration by means of a 'confirming letter'. In *Way & Waller Ltd* v *Ryde*[3] the claimants, on being instructed to find a purchaser for the defendant's hotel, wrote to the defendant enclosing a scale of remuneration which they intended to apply to this transaction The defendant did not reply to the agents' letter and, when a sale was effected, claimed that he was not bound by the enclosed scale. The trial judge held that the defendant need pay only a 'reasonable sum'; the Court of Appeal, however, took a different view. Lord Greene MR, referring to the agents' letter, said[4]

> The substantial point of it is that in that letter there is enclosed, or set out, a scale of remuneration, and the recipient of that letter must have understood that those were the terms on which the claimants were willing to work for him. With that letter before him he allowed them to continue. He allowed them to do the work ... This was the claimants' scale, be it a usual one or an unusual one, and that is the scale the defendant must be taken to have agreed to accept ... He agreed to it by

---

1   *Felthouse* v *Bindley* (1862) 11 CBNS 869.
2   A question which is considered at pp 43–46.
3   [1944] 1 All ER 9.
4   At p 10.

conduct for better or worse, and whatever it is, be it high or be it low, he is bound by it, in my opinion. That seems to me to be a perfectly clear case of a contract to be inferred from the conduct of the parties.

The 'confirming letter' which reached the courts in *John E Trinder & Partners* v *Haggis*[1] purported to alter, not the amount of commission, but the circumstances in which it should become payable. The claimants in that case were orally instructed to find a purchaser for the defendant's property; the letter which they sent to their client, however, stated that commission would be due upon the introduction of 'a person willing to sign a contract to purchase at an agreed price'. This letter was ignored by the defendant, who later disputed his liability to pay commission in respect of an introduction which did not lead to a sale. It was held by a majority of the Court of Appeal that the claimants were entitled to their commission, but a strong dissenting judgment was delivered by Denning LJ, who pointed out that the agents' letter was in law an offer and, as such, became binding upon a client only when it was accepted. Mere silence could not constitute acceptance for this purpose and, as for the client's 'conduct' in carrying on negotiations with the person whom the agents introduced, this no more suggested that he agreed to the terms of the letter than that he regarded the original instructions as still applicable.

Notwithstanding Denning LJ's note of dissent, these two cases support quite strongly the idea that an unanswered letter may be of binding force.[2] However, the matter cannot be regarded as settled, for an earlier decision of the House of Lords, which suggests the contrary, was unfortunately not cited to the Court of Appeal on either occasion. The case in question, *Toulmin* v *Millar*,[3] concerned an agent who was instructed to find a tenant for an estate. The agent sent the client a list of the charges appropriate to this work, including a note to the effect that, if a tenant who was introduced by the agent subsequently purchased the freehold, an additional commission would become payable. The client did not reply to this communication, the agent introduced a tenant, and the tenant subsequently purchased the freehold reversion. The House of Lords rejected the agent's claim for commission on this sale, on the ground that he had never been employed to find a

---

1   (1951) 158 EG 4.
2   See also the unreported case of *Britton Poole & Burns* v *Theodosiades* (1983), where the appointment of a sub-agent was held valid by virtue of a confirming letter.
3   (1887) 3 TLR 836.

purchaser for the property, but only a tenant; nor was the client bound by the agent's attempt to enlarge the scope of his authority, for this was something to which he had not positively assented.

In the light of the foregoing discussion, it must be regarded as doubtful whether the terms of a 'confirming letter' really bind a client in the legal sense. However, even if it is not actually enforceable, such a document may have a more limited but none the less useful role to play in the event of legal proceedings. After all, the question which has to be decided is upon what terms the parties in fact agreed, and a client may have acted in such a way as to indicate positive agreement with the contents of the letter.[1] Even where this is not so, a court may give some weight to the fact that the client, when faced with the agent's letter, expressed no dissatisfaction whatsoever with the terms which are now disputed.[2]

The uncertainty as to precisely what formalities are legally necessary, in order for an agent to recover commission, has now been intensified by section 18 of the Estate Agents Act 1979 and the Estate Agents (Provision of Information) Regulations 1991.[3] These statutory provisions oblige an estate agent to notify the client at the outset of their relationship of all potential charges; failure to do so may render the agent's claim unenforceable in court, whether or not the requirements of common law have been met.[4]

Remuneration where no sale results

As mentioned at the start of this chapter, an agent who works on a commission basis is not entitled to any remuneration in respect of abortive work, in the absence of a clear express provision to the contrary. As Lindley LJ explained in *Lott* v *Outhwaite*:

> It was said that there was an implied contract to pay the agent a *'quantum meruit'* for his services. The answer was that there could be no implied contract when there was an express contract.

---

1   As occurred in *Freedman* v *Union Group Ltd* [1997] EGCS 28.
2   See, for example, *Milton Marlowe & Co* v *Southcut* [1980] 1 EGLR 21; *Hampton & Sons* v *Trade & General Securities Ltd* [1979] 1 EGLR 39; *John D Wood & Co* v *Bardiger* [1986] 1 EGLR 238; *Colin Buckle & Co* v *Charterhall Properties (Chesham) Ltd* [1990] 1 EGLR 57.
3   See pp 287–294.
4   See pp 293–294.

The consequences of this were clearly spelled out by McCardie J in *Howard Houlder & Partners Ltd* v *Manx Isles Steamship Co Ltd*:[1]

> It is a settled rule for the construction of commission notes and the like documents which refer to the remuneration of an agent that a claimant cannot recover unless he shows that the conditions of the written bargain have been fulfilled. If he proves fulfilment he recovers. If not, he fails. There appears to be no halfway house, and it matters not that the claimant proves expenditure of time, money and skill.

These principles have never been seriously challenged in the case of estate agents, who recognise that, if they want any kind of fee in the event of a client's withdrawal, must make express provision for this.[2] However, the position of auctioneers has generated rather more argument, perhaps because an unsuccessful auction may result in the expenditure of considerable time, trouble and money by the auctioneer.

There is, it must be said, not much authority on the point, largely because in practice the position is usually provided for in the agreement between auctioneer and client. Thus, for example, in *Skinner* v *Andrews & Hall*[3] it was agreed that, if the property had not been sold by a certain date, there would be no charge of any description. More commonly, provision is made for the payment of a sum of money (often modest in amount), which may be inclusive or exclusive of expenses, if the sale proves abortive or the property is withdrawn by the vendor. In *Green* v *Bartlett*[4] the amount was to be £25; in *Green* v *Hall*,[5] the figure was £200 plus all advertising expenses. And, of course, where the commission agreement contains a clause similar to that used in *John Meacock & Co* v *Abrahams*,[6] it will be argued that a withdrawal of property by the vendor for the purpose of selling privately entitles the auctioneer to the full agreed commission.

Even where no express agreement has been made as to a possible failure of the sale, many vendors are willing to pay a reasonable sum

---

1 [1923] 1 KB 110, 113 (shipbroker).
2 For an example of such provision, leading to a successful claim by the agent, see *Savills Land & Property Ltd* v *Kibble* [1998] EGCS 170: p 102. Express notice to the client that such a charge may be made is also required under the Estate Agents Act 1979: see p 288.
3 (1910) 26 TLR 340.
4 (1863) 14 CBNS 681.
5 (1848) 12 LT (OS) 151.
6 [1956] 3 All ER 660: see p 134.

to the auctioneer for trouble and expense, and this too has undoubtedly helped to prevent litigation. In *Rainy* v *Vernon*[1] where, as we have seen, an auctioneer claimed to be entitled to full commission when the client withdrew property from the sale and sold it privately, the vendor was prepared to make a payment of 50 guineas for the auctioneer's services. So too, in *Frank Swain (a firm)* v *Whitfield Corporation Ltd*,[2] where the liquidator of a company withdrew an auctioneer's authority some 10 minutes before the sale was due to commence, it was agreed on all sides that the auctioneer should receive reasonable remuneration for his labours.

The few occasions on which a court has considered the position of an auctioneer, who has not in fact sold the property, have produced a remarkable diversity of judicial opinion. Some slender support for the view that a vendor who revokes the auctioneer's authority is liable to pay a reasonable sum may be found in the case of *Chinnock* v *Sainsbury*.[3] In that case an auctioneer, who had advanced money to the defendant, sued to enforce a promise to allow him to sell certain property and recoup himself from the proceeds. It was held that the promise was not of a type which equity would enforce, but that the parties were 'subject to any claim which is due to the auctioneer or agent in that character, which must be ascertained at law'.[4]

Chinnock v *Sainsbury* notwithstanding, it is well settled that an estate agent who has not succeeded in earning commission cannot recover a reasonable sum on a *quantum meruit*, and it is difficult to find any justification for treating an auctioneer any differently. In *John Meacock & Co* v *Abrahams*[5] the auctioneers, whose sale on behalf of mortgagees had been prevented by the mortgagor, claimed against their clients for commission or a reasonable remuneration. The claim for commission failed, for reasons which have already been considered.[6] The Court of Appeal was unanimous in rejecting the alternative claim, on the ground that, where an express agreement as to payment has been made, there is no room to imply a promise to pay a reasonable sum if the condition for payment is not fulfilled. Both Denning and Hodson LJJ likened the instant case to that of an estate

1   (1840) 9 C & P 559.
2   (1962) 183 EG 479.
3   (1860) 30 LJ Ch 409.
4   At p 411, *per* Romilly MR.
5   [1956] 3 All ER 660.
6   P 134.

agent, and Hodson LJ quoted with approval a passage from the speech of Lord Wright in *Luxor (Eastbourne) Ltd* v *Cooper*:[1]

> In the case of the commission agent, to whom payment is dependent on completion or the like condition, the principal does not promise that he will complete the contract for reasons which I have explained. His only promise is that he will pay commission if the contract is completed. There is no promise to pay a reasonable remuneration if the principal revokes the authority to the agent. And it is a further objection to a claim on a *quantum meruit* that the employer has not obtained any benefit. The agent has earned nothing until the event materialises.

The applicability of this principle to the case of an auctioneer was questioned by Upjohn LJ in *Frank Swain (a firm)* v *Whitfield Corporation Ltd*.[2] The matter did not, strictly speaking, arise for decision, since the client conceded that a reasonable sum was due,[3] and the sole question to be answered therefore was what amount would be 'reasonable' in the circumstances. Nevertheless, in approving the client's admission of liability (which was based on counsel's opinion), Upjohn LJ said[4]

> It is clear that an auctioneer is in rather a different position from an estate agent who contracts with his principal to introduce a purchaser. Until he introduces a purchaser who purchases, he has no right to any commission whatever; but with an auctioneer the law is different, if the properties are withdrawn from auction he is entitled to a *quantum meruit* for the services thrown away.

In the light of *John Meacock & Co* v *Abrahams*,[5] which does not appear to have been cited in *Frank Swain*'s case, it is suggested that this dictum cannot be supported. Accordingly, if an auctioneer wishes to be able to claim for time and trouble expended on an abortive sale, this intention should be made plain in the commission agreement.

---

1   [1941] AC 108, 141.
2   (1962) 183 EG 479.
3   A very questionable concession, in the light of what has been discussed.
4   (1962) 183 EG 479.
5   [1956] 3 All ER 660.

Unfair commission terms

The Unfair Terms in Consumer Contracts Regulations 1999[1] apply to any written contract under which goods or services are supplied to a consumer. They are thus potentially applicable to estate agents' standard terms of business wherever the client is a private individual (normally in residential agency).

The Regulations provide two grounds on which a contract term may be challenged. First, by virtue of regulation 5(1) a contractual term which has not been individually negotiated shall be regarded as 'unfair' if, contrary to the requirement of good faith, it causes a significant imbalance in the parties' rights and obligations arising under the contract, to the detriment of the consumer. In considering the meaning of this provision, the House of Lords in *Director General of Fair Trading* v *First National Bank plc*[2] regarded good faith as based on 'good standards of commercial morality and practice' and held that the requirement was designed to promote fair and open dealing and to prevent unfair surprise and the absence of real choice.

If unfairness in this sense is established, the term in question is not binding on the consumer, although the rest of the contract remains in force so long as it can survive without the offending term. It should, however, be noted that, so long as a term is in plain intelligible language, the assessment of its fairness shall not relate to 'the adequacy of the price or remuneration as against the services supplied in the exchange'.[3] In effect, therefore, the principle of freedom of contract applies to what might be regarded as the 'core' of the contract.

The second ground of attack is under regulation 7(1), which provides that a seller or supplier must ensure that any written term of a contract is expressed in plain, intelligible language. If this is not the case, then the term in question is to be interpreted in the way most favourable to the consumer.

In *Bairstow Eves London Central Ltd* v *Smith*,[4] the claimant agents' standard sole agency terms provided that, while the 'standard commission rate' was 3% of the sale price, a client who duly paid within 10 days of completion of the sale would qualify for an 'early

---

1 SI 1999 No 2083, replacing an earlier version (SI 1994 No 3159) in implementing EEC Council Directive 93/13.

2 [2002] 1 AC 481.

3 Regulation 6(2).

4 [2004] 2 EGLR 25.

payment discounted commission rate' of 1.5%. This provision was struck down as unfair; in the view of the court, the true position was that the commission rate was 1.5%, with a penalty of double commission in the event of late payment. This view of the realities of the situation also led the court to reject the agents' claim that there were simply two alternative commission rates and that their case was thus excluded from the statutory requirements by regulation 6(2).

Quite apart from enabling consumers to challenge unfair contract terms, the 1999 Regulations also provide for an administrative system by which such terms may be regulated. Regulation 8 gives power to the Office of Fair Trading, where it receives a complaint that a 'contract term drawn up for general use' is unfair, to apply to the courts for an injunction against any person using that term, or a term like it, or recommending such terms for use.

The case of *Office of Fair Trading* v *Foxtons Ltd*[1] concerned the exercise of this power by the OFT in respect of certain terms used by Foxtons in respect of its residential lettings business, notably those which made fresh commission payable on any renewal of a tenancy or on a sale by the landlord to the tenant, and also a provision that the original client would remain liable for such fees even after he or she had disposed of the property. In dealing with a preliminary issue in this litigation, the court recognised an essential difference between a challenge by an individual consumer and one by the OFT. In the former case, it was said, that court must consider all the circumstances of the individual transaction before deciding whether or not the term was 'unfair'. In the case of a 'collective' challenge by the OFT, on the other hand, the question of 'fairness' could only be tested by reference to a hypothetical "typical consumer".

In the eyes of the trial judge,[2] the consequence of this analysis was that, while the OFT might be entitled to an injunction preventing the future use of these terms, it could not prevent the agents from relying on these terms in contracts already made; this was because the agents might be able to show that, in the circumstances of an individual case, the term was not unfair. Nor was the OFT entitled to a declaration that any particular term was necessarily 'unfair', since this too would depend on all the facts of a particular case.

While accepting the basic distinction between the two types of legal challenge to unfair terms, the Court of Appeal unanimously

---

1   [2009] EWCA Civ 288.
2   [2008] EWHC 1662 (Ch).

rejected the judge's conclusions. Emphasising that the whole purpose of the legislation is consumer protection, and arguing that the system of pre-emptive challenges by the OFT was a more effective means of achieving this aim than leaving matters to be raised by individual consumers, the court held that it was necessary to allow the OFT to seek an injunction covering existing contracts as well as future ones. Similar reasoning led to a ruling that the OFT could indeed seek a declaration on the unfairness of particular terms.

Expenses

As a general rule, every agent has a right to be indemnified by the principal against all losses, liabilities and expenses incurred in the lawful performance of the agent's duties. However, while an estate agent may undoubtedly claim an indemnity where carrying out a client's instructions results in legal liability, it is presumed that the agent's expenses will be covered by the commission which is earned on a successful transaction. In consequence, an estate agent who wishes to ensure that expenses will be met, whether or not a sale results, must provide expressly for this in the terms of appointment.[1] Moreover, section 18 of the Estate Agents Act 1979 requires such express provision to give a detailed breakdown of the charges for which the client may become liable.[2]

Auctioneers, like estate agents, cannot recover their advertising and other out-of-pocket expenses in addition to the commission which is payable on a sale of the property, unless their terms of business make specific provision to this effect. As with all commission agents, their agreed fee is presumed to be inclusive of expenses.[3] Notwithstanding this general principle, however, there is at least some authority for the view that an auctioneer who fails to earn any commission (because no sale has resulted) may have a legitimate claim against the client for expenses incurred. This is based partly on the *dictum* of Romilly MR in *Chinnock* v *Sainsbury*,[4] to which reference has already been made.[5]

---

1   For a successful claim based on such a provision, see *Bernard Thorpe & Partners* v *Flannery* (1977) 244 EG 129.
2   See p 288.
3   *Marshall* v *Parsons* (1841) 9 C & P 656.
4   (1860) 30 LJ Ch 409, 411.
5   P 146.

Indeed, in one report of that case,[1] the Master of the Rolls is said to have referred expressly to the auctioneer's expenses. There is also a dictum of Denning LJ in *John Meacock & Co v Abrahams*,[2] to the effect that, while the auctioneers were not entitled to any fees on a *quantum meruit* basis, they would have been entitled to their expenses.

Despite this slender authority, a claim that auctioneers are to be treated differently from other commission agents seems an inherently weak one, and prudent auctioneers will ensure that the position in respect of expenses is specifically dealt with in their agreement with the client.[3] Indeed, the majority of auctioneers do precisely that, which is presumably why this issue has so seldom reached the courts.

Indemnity

In principle, every agency relationship contains by implication a promise on the part of the principal to indemnify the agent against any legal liability to which the agent may become subject as a result of carrying out the principal's instructions. As Vaughan Williams LJ said in *Williams v Lister & Co; Llewellyn Bros, Third Parties*:[4]

> The law does imply an undertaking to indemnify the agent for injury incurred in the course of carrying out his duty as agent.

An auctioneer who receives, as stakeholder, a deposit from the highest bidder, will incur personal liability for handing this over to the vendor before the latter becomes legally entitled to it.[5] Should this occur, so that the auctioneer has to pay the amount of the deposit to the purchaser, the auctioneer can then recover this sum from the vendor under the indemnity principle. Furthermore, if an auctioneer who is still holding the deposit is successfully sued by the purchaser for its return, the vendor is liable to indemnify the auctioneer against the legal costs incurred.[6]

Other examples of the right to indemnity include *Brittain v Lloyd*,[7] where an auctioneer recovered from the vendor the amount of auction

---

1   (1860) 3 LT 258, 259.
2   [1956] 3 All ER 660.
3   This is also a requirement under the Estate Agents Act 1979; p 288.
4   (1913) 109 LT 699, 700.
5   See p 245.
6   *Spurrier v Elderton* (1803) 5 Esp 1.
7   (1845) 14 M & W 762.

duty which the Commissioners of Excise required the auctioneer to pay on property bought in. And in *Warlow* v *Harrison*,[1] where the Exchequer Chamber held that an action would lie against an auctioneer for failing to knock down property which was advertised as 'without reserve', Martin B said:[2]

> If the auctioneer had contracted any liability in consequence of his employment and the subsequent revocation or conduct of the owner, he is entitled to be indemnified.

Lien

There appears to be no direct authority on the question whether an estate agent may reinforce a claim to commission or expenses by exercising a lien over any property or money belonging to the client which comes into the agent's possession by virtue of the agency. However, it is submitted on principle that there is no reason to treat estate agents any differently from other categories of agent, such as auctioneers, in this respect. If this is correct, it means that an estate agent is entitled to exercise a particular lien, that is, to retain possession of the client's goods or money until all charges in respect of those particular items have been met. In practice this right will only be of value to an agent who is given possession of title deeds or who is still holding a deposit at a stage in the transaction when it belongs to the client;[3] until that time there is nothing belonging to the client to which the agent's lien may attach.

The question of an estate agent's lien may be a matter of speculation, but there is clear authority for the proposition that an auctioneer may exercise a lien over the client's goods or the proceeds of sale. The case of *Palmer* v *Day & Sons*[4] indicates that such a lien is a particular one, which does not extend to the general balance of account between the parties. The defendants there, who had sold a house and certain furniture for a client, were authorised by his trustee in bankruptcy to sell some pictures on behalf of the estate. The

---

1   (1859) 1 E & E 295: p 180.
2   At p 317.
3   This will not usually occur until completion of the sale: see *Skinner* v *Trustee of the Property of Reed* [1967] Ch 1194.
4   [1895] 2 QB 618; *Dirks* v *Richards* (1842) 5 Sco NR 534. See also *Webb* v *Smith* (1885) 30 ChD 192: p 153.

defendants did so, retaining out of the proceeds of this sale their charges in respect of the earlier ones. It was held, on a technical point of bankruptcy law[1] that the auctioneers were entitled to do so, but the Divisional Court of Queen's Bench made it plain that the auctioneer's lien could not in itself have justified this action.

The case of *Webb* v *Smith*[2] offers a good illustration of an auctioneer's lien in operation. In that case the defendant auctioneer sold a brewery on behalf of a client, retaining part of the proceeds of sale. The defendant had previously sold some furniture for the same client, and was also still holding the balance of this purchase money. The claimant, to whom the client was indebted, claimed to have a charge over the brewery proceeds, and this was acknowledged by the auctioneer; despite this acknowledgment, however, the auctioneer handed over to the client the proceeds of the sale of the furniture. It was held that the auctioneer was still entitled, as against the claimant, to a lien over the proceeds of sale of the brewery; he could not be compelled to take his payment from the other fund, over which he had no lien.

The purpose of the auctioneer's lien is to assist in the enforcement of rights against the vendor; consequently the lien only attaches to goods and money which are the property of the vendor. This was made clear in the case of *Skinner* v *Trustee of the Property of Reed*[3] which concerned the sale by auction of a farm. The farm, which was mortgaged, was stated to be sold free from encumbrances. It was knocked down at the reserve price of £30,000, and a deposit of £3,000 was paid to the auctioneer as stakeholder. Upon completion of the contract it was discovered that the amounts outstanding upon the mortgages were greater in total than the purchase price. Since the vendor was bankrupt, the purchaser called upon the auctioneers to apply the whole of the deposit towards discharging the mortgages. Cross J held that the auctioneers must comply with this demand; the deposit money had never become the property of the vendor, as it was required to discharge encumbrances according to the conditions of sale; consequently there was nothing to which the auctioneer's lien could attach.

---

1   Bankruptcy Act 1883, s 38. See now the Bankruptcy Act 1914, s 3.
2   (1885) 30 ChD 192.
3   [1967] Ch 1194.

The Contract of Sale

A client who instructs an estate agent or auctioneer to market property almost invariably does so in the hope that the services provided will culminate in a sale or lease of that property. However, while the bringing about of a contract of sale or lease may be regarded as the core function of both estate agents and auctioneers, there is an important difference between them. In normal circumstances, the activities of an auctioneer will be sufficient in themselves to bring about a binding contract, but those of an estate agent are of more limited effect: the agent's task is merely to introduce a potential purchaser or tenant who will, following further negotiations with the client (which may themselves be conducted through the agent) put in motion the normal conveyancing procedures by which contracts for the sale of land are created.[1]

In this chapter we deal with a number of separate matters related to the making of a contract of sale or lease. First, we consider the (very limited) extent to which the law will regard an estate agent, in the absence of clear instructions, as having implied authority to commit the client to such a contract. Second, we examine the implied authority of an auctioneer, which, not surprisingly, is greater in this respect. Finally, we review a number of special rules governing the way in which the process of bidding at auction actually brings about a binding contract, and the various pitfalls by which the path to that contract is beset.

---

1. In the vast majority of cases, this will involve a so-called 'exchange of contracts' by the parties' solicitors.

In order to make sense of the decided cases in this area, it is important to appreciate that many of them were decided at a time when the legal formalities required for making a valid contract were significantly different from those which apply today. Briefly, the law did not require a contract for the sale of either land or goods to be made in writing; however, such a contract was not enforceable by a court unless the claimant could produce a document containing written evidence of its terms, signed by or on behalf of the defendant.

In the context of our present discussion, these legal rules had two very important implications. First, the evidence needed to support a contract for the sale of land by private treaty could be validly provided by an exchange of letters between the parties or their authorised agents, without the need for a formal 'exchange of contracts' in the modern sense. This raised the possibility that an estate agent, by omitting the hallowed phrase 'subject to contract' from a letter to a purchaser, might inadvertently provide the necessary written evidence to commit the vendor to a contract of sale. Inevitably, attention would then turn to the question whether the agent had any kind of authority from the vendor to make the alleged contract.

Secondly, the requirement of written evidence meant that a contract made at auction, though legally valid, remained unenforceable unless and until a memorandum of sale was drawn up and signed by or on behalf of the parties. Not surprisingly, therefore, the attention of the courts was frequently directed more at the memorandum (and at such questions as whether or not the auctioneer could validly sign on behalf of either or both parties) than at the making of the contract itself.

Both these issues fall to be viewed rather differently, in the light of legislative changes to the underlying legal rules. First, the formalities required for a sale of land by private treaty were significantly tightened up by section 2 of the Law of Property (Miscellaneous Provisions) Act 1989. This provides that, with certain exceptions,[2] a contract for the sale or other disposition of an interest in land can only be made in one of two ways: there must be either a single document containing all the contract terms, which is signed by or on behalf of both parties, or two documents, each containing all the contract terms, each signed by one party and then exchanged. It seems inherently

---

1   Law of Property Act 1925, section 40 (sales of land); Sale of Goods Act 1893, section 4 (sales of goods).

2   Notably a 'contract made in the course of a public auction' (see below).

mplausible that an estate agent could inadvertently satisfy these requirements, in the way that occurred in a number of the early cases. In consequence, the question of the agent's authority to commit the client to a contract is today likely to arise only in cases where the agent has, for whatever reason, genuinely attempted to bring about a binding agreement between the parties.

The second important legal development is that, where property is sold by auction, written evidence is no longer required to make the sale enforceable. In respect of sales of goods (whether by auction or otherwise) the requirement was abandoned in 1954.[1] As for sales of land, the Law Reform (Miscellaneous Provisions) Act 1989, while repealing section 40 of the Law of Property Act 1925, provides specifically that the stringent new rules described above do not apply to 'a contract made in the course of a public auction'.[2] It follows that those consequences which were formerly regarded as depending on whether or not a memorandum of sale had been drawn up and signed have today ceased to be of legal relevance.

Authority of estate agent

The general rule

The English courts have long held the view that the sale of land, with all its accompanying technicalities, is not something to be entered into lightly and without proper consideration. This attitude has found expression in, among other things, a reluctance to construe informal or casual agreements as intended to be legally binding (hence the eagerness with which such phrases as 'subject to contract' are seized upon). More important in the present context, it has led to a requirement that any person claiming to sell land on behalf of another must show clear evidence of authority to do so.[3]

An early manifestation of this latter requirement is found in the case of *Godwin* v *Brind*,[4] where the five joint owners of an estate

---

1   Law Reform (Enforcement of Contracts) Act 1954, section 2, repealing the Sale of Goods Act 1893, section 4.
2   Law of Property (Miscellaneous Provisions) Act 1989, sections 2, 4 and Schedule 2.
3   Even a solicitor has no implied authority to exchange contracts on a client's behalf: see *Suleman* v *Shahsavari* [1989] 2 All ER 460.
4   (1868) LR 5 CP 299n.

advertised it for sale and named Benjamin Francis, one of the five, as a person to whom anyone interested could make application 'to treat and view'. Francis purported to make a contract for the sale of the property to the claimant who, when the other four owners denied liability brought an action against them. It was held that, notwithstanding the definition of 'treat' in a leading dictionary as 'negotiate, settle, come to terms', Francis had no authority to conclude a contract on behalf of his co-owners. Two years later the disappointed claimant succeeded in recouping his loss from Francis himself, in an action for breach of warranty of authority.[1]

The first reported case dealing with the authority of an estate agent as such was *Hamer* v *Sharp*,[2] where an agent was requested to 'procure a purchaser' for certain property at a price of £2,800. The agent accepted a deposit of £280 from the claimant and signed a receipt which described the money as paid 'on account of purchase'. The corresponding acknowledgement signed by the claimant stipulated a date for completion of the contract and stated that the vendor was to supply an abstract of title; it did not, however, contain any special conditions as to the title and to this extent, therefore, it was an 'open contract'. It was held that the making of such a contract was beyond the scope of the estate agent's authority, so that his client was not legally bound by it. Hall V-C, concerned at the burdens placed upon a vendor of land by open contracts, said of the agent: 'If he had a right to enter into any contract at all, it was one of a different description.'

The actual decision in *Hamer* v *Sharp* is thus of fairly limited application. A comment of the judge has, however, been used to support a much more general proposition:

> I state my opinion to be, that when instructions are given to an agent to find a purchaser of landed property, he, not being instructed as to the conditions to be inserted into the contract as to title, is not authorised to sign a contract on the part of the vendor.

Four years later, in the case of *Wilde* v *Watson*,[3] an Irish court based its decision firmly upon the general principle expounded in *Hamer* v *Sharp* In holding that an estate agent, who had been instructed to procure an offer of £2,900 by a certain date, had no authority to make an open

---

1   *Godwin* v *Francis* (1870) LR 5 CP 295.
2   (1874) LR 19 Eq 108.
3   (1878) 1 LR Ir 402.

contract of sale, the judge said that, were this not so, the common practice of instructing a number of agents could lead to the owner becoming bound to sell to several different parties at the same time.

The opinion of Hall V-C quoted above was enthusiastically endorsed by Kekewich J in two cases tried in the Chancery Division of the High Court. In *Prior* v *Moore*,[1] an estate agent was instructed to place certain property on his books for sale and informed that the lowest price which would be acceptable to the vendor was £1,600. In due course the estate agent purported to sell the property and the purchaser sued for specific performance of the contract. The actual decision was that the agent had no authority to make the contract in question (it is not clear from the report what sort of contract this was) but the judge went on to cast doubt on the apparent 'loophole' left by Hall V-C in *Hamer* v *Sharp*, that is, the insinuation that some contracts might be within the scope of the agent's authority: 'His Lordship was sorry the Vice-Chancellor did not state what contract the agent was entitled to enter into. The general result seemed to be that the agent was not entitled to enter into any contract.'

In *Chadburn* v *Moore*,[2] an English court was able for the first time to reach a decision based on an estate agent's lack of any implied authority to contract on behalf of a client. The defendant there instructed a firm of estate agents to find a purchaser for leasehold property consisting of 34 houses. On receiving an offer from the claimant, the owner instructed the agents to withdraw five of the houses, and named the lowest price which he would accept for the remainder. It was held that the agents had no authority to accept the offer which the claimant then made and that the defendant was accordingly not bound by the open contract which they had purported to make. Kekewich J said:

> Unless express authority is given to the agent to sell, and for that purpose to enter into a binding contract, the principal reserves his final right to accept or refuse.

After this decision, the courts appeared to accept the general principle as valid and no longer found it necessary to inquire into the type of contract which the estate agent had purported to make. The few cases

---

1   (1887) 3 TLR 624.
2   (1892) 61 LJ Ch 674.

in which a vendor was held to be bound turned very much on their own peculiar facts[1] and, in *Lewcock* v *Bromley*,[2] Sargant J was able to say:

> The law on this point is quite settled that a general authority to an agent to find a purchaser does not authorise the agent to sign a contract binding on the vendor.

All the cases so far mentioned concerned the sale of property by an estate agent, but it should be noted that the same principle applies to the creation of a lease. In *Thuman* v *Best*[3] the defendant put his flat into the hands of a number of estate agents and one of these introduced the claimant, who expressed himself willing to pay the required rent of £195 per year for 'a usual West End lease'. The estate agent purported to accept this offer and the claimant brought an action against the owner for specific performance. It was held that the alleged agreement could not possibly be enforced since, even if it did not fail for uncertainty, it lacked the written evidence required by statute. Furthermore, the estate agent had, by purporting to sign a contract on behalf of his client, exceeded his authority, which was no greater in the case of a lease than that of a sale. Parker J stated with admirable clarity and concision the judicial view of an estate agent's function:

> Estate agents as such have no general authority to enter into contracts for their employers. Their business is to find offers and submit them to their employers for acceptance.

Instructions to sell

Where what is in issue is the express authority given to an estate agent by a client, it seems from various judicial remarks that, if this is to include the making of a binding contract, very clear evidence will be needed. In order to decide whether the agent has been authorised to do more than carry out the normal function of effecting introductions, the exact instructions given by the client must be ascertained, and the court is then faced with the task of interpreting these instructions in accordance with past cases and in the light of any previous course of dealing between the parties.

---

1   See, for example, *Allen* v *Whiteman* (1920) 89 LJ Ch 534.
2   (1920) 127 LT 116.
3   (1907) 97 LT 239. See also *Yallop* v *Fosh* (1953) 161 EG 603.

The cases already considered show categorically that the general rule is not displaced by instructing the estate agent to 'find a purchaser' or to 'procure an offer'. Moreover, this applies even where the client mentions a specific price at which he or she is prepared to settle. In *Chadburn* v *Moore*[1] the instructions relating to the property in its revised form were 'to find a purchaser and to negotiate a sale', and again the estate agent was held to have no more authority than usual.

The Irish case of *Carney* v *Fair*[2] shows the reluctance of the courts to infer authority from any save the most unequivocal of instructions. In that case a land agent, who looked after certain estates for the defendant, was approached by the claimant with a view to buying the land. The agent wrote to the defendant's solicitors saying he 'had reason to believe that he could secure a purchaser at £6,000 cash' and the solicitors, having taken instruction, replied that the defendant 'will accept a purchaser at £6,000'. Once again the court could find no reason to depart from the general rule expressed in *Hamer* v *Sharp*, and the claimant was held unable to enforce the contract alleged to have been made with the land agent.

Conflicting expressions of judicial opinion have been provoked by the simple instruction to 'sell' property, and the legal effect of such a request remains unclear. In *Chadburn* v *Moore*,[3] Kekewich J was of the opinion that an estate agent obtained no extra authority from being instructed to sell, unless the agent was also expressly authorised to enter into a binding contract. This view was endorsed in *Lewcock* v *Bromley*,[4] where Sargant J said: 'There must, to justify such a signing, be a special and express authority to sign'.

Sargant J treated as depending on its own special circumstances the decision of Buckley J in *Rosenbaum* v *Belson*,[5] where a client requested a firm of estate agents in writing: 'Please sell for me my houses ...' and named a price of £800. The estate agents obtained an offer of £785 which the client agreed to accept and, on the facts of the case, the estate agent was clearly empowered to sign the open contract in question. However, Buckley J, after reviewing all the previous English cases, came to the conclusion that, even without the specific consent of the client, the instruction to 'sell' would still have

---

1   (1892) 61 LJ Ch 674.
2   (1920) 54 Ir LT 61.
3   (1892) 61 LJ Ch 674.
4   (1920) 127 LT 116.
5   [1920] 2 Ch 267.

empowered the agent to make a contract. As the judge remarked: 'A sale *prima facie* means a sale effectual in point of law, including the execution of a contract where the law requires a contract in writing.' Noting that the authority in *Hamer* v *Sharp* was to 'find a purchaser', Buckley J said:

> To my mind there is a substantial difference between those expressions Authorising a man to sell means an authority to conclude a sale; authorising him to find a purchaser means less than that — it means to find a man willing to become a purchaser, not to find him and also make him a purchaser.

The effect of an authority to sell property arose again in the case of *Keen* v *Mear*[1] which is discussed in more detail below. Having there considered a number of cases concerning the estate agent's authority, Russell J said:

> The result of those cases is, in my opinion, this: that the mere employment by an owner of an estate agent to dispose of a house confers no authority to make a contract; the agent is solely employed to find persons to negotiate with the owner; but, if the agent is definitely instructed to sell at a defined price, those instructions involve authority to make a binding contract and to sign an agreement.

Had the matter rested there, it might well have appeared that the opinions expressed in *Rosenbaum* v *Belson* and *Keen* v *Mear*, coming after a survey of precedent, were preferable to the contrary opinions shown in *Chadburn* v *Moore* and *Lewcock* v *Bromley*. However, in *Wragg* v *Lovett*[2] the Court of Appeal had its first opportunity to express a view upon the question. On the facts of the case, the firm of estate agents were held to have been given express authority to do whatever was best for their client, including making a binding contract on his behalf. Thus the following remarks of Lord Greene MR (who delivered the judgment of the court) were strictly *obiter*:[3]

> We must not be understood as suggesting that when a vendor merely authorises a house agent to 'sell' at a stated priced he must be taken to be authorising the agent to do more than agree with an intending purchaser

---

1   [1902] 2 Ch 574.
2   [1948] 2 All ER 968.
3   The passage was none the less approved by the Privy Council in *Jawara* v *Gambia Airways* [1992] EGCS 54.

the essential (and, generally, the most essential) term, ie, the price. The making of a contract is no part of an estate agent's business, and, although, on the facts of an individual case, the person who employs him may authorise him to make a contract, such an authorisation is not lightly to be inferred from vague or ambiguous language.

The effect of this *dictum* is to confuse still further the legal effect of the instruction to 'sell' property. However, the practical lesson to be drawn from the cases is obvious (especially in view of the disappointed purchaser's right of action against the agent for breach of warranty of authority):[1] unless the client's written instructions clearly and unequivocally empower the estate agent to sign a binding contract of sale, the agent should go no further than introducing prospective purchasers to the client.

As a postscript to this discussion, it may be noted that, even where express authority does exist, any sale which is effected by the agent must be in strict accordance with its terms. Thus, in the New Zealand case of *Ludwig* v *Schultze*,[2] where an estate agent was clearly instructed to effect a binding contract of sale, the client was none the less held not to be bound by a sale at a lower price than had been authorised. It was irrelevant that the purchaser knew nothing of the restriction on the agent's authority; there was no implied authority in this situation, so the express terms were all that mattered.

Terms of sale

It is clear, from what has been said above, that an estate agent may be given authority to enter into a contract of sale or lease on behalf of the client. Where, however, the court decides that such authority has been given, a further problem of interpretation remains to be solved. The question which must be asked is this: do the instructions under which the agent is acting empower the making of any sort of contract at all, or is the agent restricted to making an 'open contract', that is, one which leaves all the terms except the price to be implied by law? In particular, would an agent acting under general instructions 'to sell and exchange contracts' be permitted to make use, for example, of the National Conditions of Sale, a standard form document containing most of the detailed conditions subject to which land is commonly sold?

---

1   As in *Godwin* v *Francis* (1870) LR 5 CP 295.
2   (1885) NZLR 4 SC 247.

When the courts were first called upon to consider the question of the general authority of estate agents, the conditions which were implied into open contracts by common law placed the vendor under onerous obligations, especially in connection with the satisfactory proof of title. Vendors were accordingly very reluctant to enter into open contracts for the sale of their property, preferring to modify their common law obligations by inserting into the contract of sale express terms, drafted for the purpose by their legal advisers. In *Hamer* v *Sharp*,[1] where an estate agent purported to make an open contract, the actual decision was that he had no implied authority to make such a contract. So too, in *Wilde* v *Watson*,[2] although the decision rested upon a broader principle, the judge was clearly concerned at the idea of the vendor being bound by an open contract. Indeed, considerable weight was attached in that case to the fact that the agent still had in his possession the conditions under which the property was to have been sold at auction; this should have made him more careful to guard against committing his client to anything more burdensome.

The Real Property and Conveyancing Act 1881 greatly improved the position of vendors under open contracts (particularly where leasehold property was being sold) and thus made the use of express conditions less necessary. Although this alteration in the law failed to convince Kekewich J in *Chadburn* v *Moore*[3] that estate agents should now be presumed to have authority to contract, it nevertheless led to a change of judicial attitude regarding the type of contract which might be entrusted to such agents. Hitherto, the protection of the client from the rigours of an open contract had been of paramount importance; henceforth, more weight could be given to another judicial view, that the drafting of conditions of sale should be left to solicitors. Significantly, in *Saunders* v *Dence*,[4] the first case in which an estate agent was held to have been given authority to make a binding contract on behalf of a client, no special drafting was necessary. In that case the defendant, who had bought a farm at an auction for £6,000, authorised the auctioneer to transfer his interest for £6,600. In holding that this empowered the auctioneer to sign a contract, Field J drew attention to the fact that the property was sold on exactly the same conditions as those on which the defendant had bought it; thus, no complications arose as to title.

---

1   (1874) LR 19 Eq 108.
2   (1878) 1 LR Ir 402.
3   (1892) 61 LJ Ch 674.
4   (1885) 52 LT 644.

The question of the extent of an estate agent's authority arose specifically in *Keen* v *Mear*,[1] which concerned a cottage owned by two brothers, Samuel and Walter Mear. Samuel authorised an estate agent named Cox to 'sell' the cottage and Cox, having accepted a deposit from the claimant, signed a contract which set a date for completion and contained conditions relating to the title. In considering the effect of the agent's instructions, Russell J said:

> In these circumstances, Cox had, in my opinion, authority to sign an open contract on behalf of Samuel Mear to sell the property for £515. But had he authority to sign the particular contract here in question? I think not. It contains a special condition as to title, which might or might not be less favourable to a vendor than the title required under an open contract.

The agent had therefore exceeded his authority in making this contract. However it was held that, on the facts, Samuel had ratified the agent's act, thus giving him authority in retrospect and making Samuel liable on the contract. The result of the case was that Samuel had to pay damages to the claimant; specific performance was out of the question, since the whole transaction had taken place without the consent, or indeed the knowledge, of Walter Mear.

The quotation from the judgment of Russell J given above may appear to suggest that, if the special condition in question could be shown conclusively to be for the benefit of the vendor, an estate agent would have implied authority to include it. This was the view taken by the trial judge in *Wragg* v *Lovett*;[2] the Court of Appeal, however, concerned at the difficulty of weighing the relative merits and demerits of each individual condition, interpreted this *dictum* rather differently, saying (i) that the words of Russell J:

> suggest to our minds that he meant to limit the authorisation in such a case to an authorisation to make an open contract and to exclude a consideration whether a term inappropriate to an open contract was or was not more beneficial to the vendor; (ii) that an open contract is a thing certain, whereas the question whether or not a particular stipulation is more or less beneficial to the vendor may, in the complications of English conveyancing, be a question of great difficulty which could only be resolved by litigation.

---

1   [1920] 2 Ch 574.
2   [1948] 2 All ER 968.

The Court of Appeal was at pains to point out that no concluded opinion was necessary on this matter since, on the special facts of the case, the estate agents had been given express authority to enter into a contract of considerable complexity. Nevertheless, it would be a foolhardy estate agent who, in the teeth of such judicial remarks, relied on such general words of authority as 'sell' to make anything other than an open contract.

Special cases

It is established, then, beyond any reasonable doubt that the mere placing of property in the hands of an estate agent does not endow that agent with any implied authority to make a contract for the sale or other disposition of the property.[1] This is not to say, however, that authority to that effect can never be conferred upon the agent; what it means is that it must take one of the other forms which we discussed in Chapter 1. In the Irish case of *Brennan* v *O'Connell*,[2] for example, an estate agent telephoned clients to tell them that he had signed a contract for the sale of their farm, whereupon they expressed unqualified approval of his (unauthorised) action. However, the agent did not tell the clients that another prospective purchaser had shown interest in the property and, as a result, when the agent's purchaser sought to enforce the contract, the clients denied that they were bound by it. This was on the ground that their lack of knowledge precluded a finding that they had ratified the agent's act.[3] The Irish Supreme Court, however, held that the fact of which the clients had been kept in ignorance was not, objectively speaking, sufficiently material to obviate what appeared to be a clear case of ratification; the sale was therefore binding.

In *Worboys* v *Carter*[4] a land agent representing the tenant of an agricultural holding believed that his client, who was serving a short term of imprisonment, had authorised him to assign the tenancy to the claimant. This was not in fact true but the defendant client, on being released from prison, allowed the claimant to believe that the contract

---

1   The decision of the Irish Supreme Court in *Kelly* v *Park Hall Schools Ltd* (1978) 113 ILTR 9 can only be justified on the basis that the agent was expressly authorised to sell.

2   [1980] IR 13. See also *Keen* v *Mear* [1920] 2 Ch 574.

3   See p 16.

4   [1987] 2 EGLR 1.

of assignment was valid and binding. The Court of Appeal was not prepared to follow the trial judge in holding that the defendant had ratified his agent's act (since the evidence showed that he was in truth always opposed to what had been done). However, the court held that the claimant was none the less entitled to enforce the contract of assignment, on the basis that the defendant was estopped by his conduct from disputing its validity.[1]

The doctrine of apparent authority may well serve to bind a client in this context; a good illustration, and one with serious implications for practising estate agents, is *Walsh* v *Griffiths-Jones*.[2] The claimant there, who owned a house divided into three flats, wished these to be occupied pending the sale of the whole property, but did not want any occupier to acquire a protected tenancy. The claimant accordingly instructed an estate agent (one who had advertised to the effect that he could avoid the provisions of the Rent Act) to grant a licence, but not a lease, of each of the flats to a suitable applicant. When the defendants went to view one of the flats in question, they met the claimant herself, who told them that the agent would attend to all the details of the agreement. The defendants duly visited the agent and entered into a form of licence agreement which, it was subsequently held, was a sham and consequently took effect as a lease within the Rent Act. Notwithstanding this finding, the claimant argued that she was not bound, since the agent had exceeded his authority. However, the county court judge, who described the case as 'one of the clearest cases of holding out an agent as having full authority to enter into a contractual agreement that I can recall', held that, whatever might be the legal effect of the agreement between the defendants and the agent, the claimant was bound by it.

Authority of auctioneer

It is very clear from the above discussion that, in the eyes of the English courts, the primary function of an estate agent is merely to introduce a prospective purchaser to the client. By contrast, an auctioneer is normally expected to bring about a legally binding contract for the sale of the client's property and will therefore be specifically instructed to

---

1   Following *Spiro* v *Lintern* [1973] 3 All ER 319: p 13.
2   [1978] 2 All ER 1002. See also *Hayes* v *Douglas* [1976] 5 WWR 308.

sell. Even without such specific instructions, however, it seems that authority to carry out a sale is implicit in the nature of the profession. As Lord Ellenborough CJ has said:[1]

> If the principal send his commodity to a place, where it is the ordinary business of the person to whom it is confided to sell, it must be intended that the commodity was sent thither for the purpose of sale ... If one send goods to an auction-room, can it be supposed that he sent them thither merely for safe custody? Where the commodity is sent in such a way and to such a place as to exhibit an apparent purpose of sale, the principal will be bound, and the purchaser safe.

In the situation here envisaged, a purchaser will obtain good title, not only where nothing has been said to the auctioneer, but even where the vendor has strictly forbidden a sale. The auctioneer, however, will of course be liable to the vendor for any disobedience of express instructions.

Sale by private treaty

It is important to note that an agent's implied authority to sell (and, for that matter, any express authority which is given in general terms) is not unlimited in scope. The auctioneer is authorised to carry out a sale in accordance with normal trade practice, having regard to the subject-matter of the sale. In particular, an auctioneer has no implied authority to sell by private treaty; unless express instructions to the contrary are given, the sale must be by auction. In *Daniel* v *Adams*[2] the defendant and his wife, who were joint owners of two houses, instructed their steward to sell these by public auction. Two days later, the defendant wrote to the steward, stating that he would not accept less than £120 for the property. The steward, believing that he was authorised to sell, agreed to sell the houses to the claimant for £150. It was held that the wife, who had not joined in writing the letter, could only be bound by acts within the steward's original authority, which did not include a private sale.

The principle embodied in this case was applied to an auctioneer as such in *Marsh* v *Jelf*[3] where the claimant auctioneer, who had been employed to sell the defendant's property, bought it in at the reserve price. One month later, the property was sold to a person who had

---

1   *Pickering* v *Busk* (1812) 15 East 38, 43.
2   (1764) Ambl 495.
3   (1862) 3 F&F 234.

attended the auction and the auctioneer claimed commission for negotiating the sale. It was held that, even if this private sale had been negotiated by the auctioneer (and not, as the defendant contended, by his solicitor) he would not be entitled to any payment for this act, since it lay outside the scope of his authority. Moreover, the auctioneer was not permitted to bring evidence of a custom among auctioneers to charge commission in such cases, since a trade custom of this type would not bind the general public.

Although it is clear from the foregoing that the mere placing of property in the hands of an auctioneer does not impliedly authorise a sale other than by auction, a wider authority is frequently contained in the client's instructions.[1] Indeed, the circumstances of a particular case may show that the normal principle has been displaced. In *Else* v *Barnard*[2] certain property was ordered by the court to be sold by auction in two lots. The auctioneer was given a sealed envelope containing a reserve price, which he was to open at the time of the sale. In the event of bidding failing to reach the reserve, the auctioneer was instructed merely to buy it in without revealing the reserve price, and this is what happened to the first lot. The successful bidder for the second lot, having discovered from the auctioneer's clerk the amount of the secret reserve upon lot 1, offered this sum to the auctioneer, who was still on the rostrum. The auctioneer agreed and Courtauld signed a bidding paper for the two lots. Courtauld later regretted his action, but the court held him bound by the contract. As Sir John Romilly MR pointed out, the bidding paper clearly treated this as a sale by auction; further, there was no reason why the purchaser should not agree to pay the reserve price, despite the curious secrecy over the figure.

A similar conclusion was reached in *Bousfield* v *Hodges*,[3] where a legal action concerning a deceased person's estate was compromised on the terms that it should be sold by public auction and the proceeds divided. The auction having proved abortive, the auctioneer sold privately for the amount of the reserve price to the defendant, who later refused to proceed with the purchase, claiming that the auctioneer had no authority to sell by private treaty. It was held, however, that the clear intention of the vendors was to sell, provided only that the reserve price was obtained; consequently the defendant was bound to complete.

---

1   See, for example, *Green* v *Bartlett* (1863) 14 CBNS 681, where the auctioneer was authorised to sell 'by public auction or otherwise'.

2   (1860) 28 Beav 228.

3   (1863) 33 Beav 90.

In both *Else* v *Barnard* and *Bousfield* v *Hodges*, where private sales were held to have been impliedly authorised, the vendors raised no objection to what had been done on their behalf; indeed, in each case it was the vendor who sought to enforce the sale. These two cases might equally therefore have turned upon a ratification by the vendor of an unauthorised act by the auctioneer. However, in *Garnier* v *Bruntlett*,[1] where an auctioneer sold a farm privately at its reserve price immediately after an abortive auction, it was the clients who sought to resile. After a detailed consideration of the evidence, Whitford J concluded that the clients' instructions, as communicated through their solicitors to the auctioneer, were that the property should be sold; the method of sale was of secondary importance. As a result, the sale was held to be binding. Nevertheless, the decision was a close one, and it is suggested that a prudent auctioneer should always obtain express authorisation from the client before entering into a private contract of sale.

Terms of sale

A contract of sale entered into at auction must, if it is to be enforced against the vendor, be in accordance with the terms which the vendor has laid down.[2] Thus, it appears, the auctioneer has no implied authority to vary those terms.[3] In *Moore* v *Khan-Ghauri*,[4] a purchaser claimed to have been promised by the auctioneer that the fall of the hammer would be merely conditional and that he would be given the opportunity to inspect the property. The Court of Appeal swiftly rejected this claim, holding that an auctioneer would have no authority to bind the vendor by such an undertaking.

One situation in which this problem surfaces is where inaccuracies in the printed sale particulars are corrected by the auctioneer from the rostrum. In *Manser* v *Back*,[5] for example, the

---

1   (1974) 236 EG 867.
2   See the judgment of Hullock B in *Jones* v *Nanney* (1824) 13 Price 76, 110.
3   In *D & M Trailers (Halifax) Ltd* v *Stirling* [1978] RTR 468, where an auctioneer sold
     a vehicle by private treaty after an auction, it was held by the Court of Appeal that
     an exemption clause which would have formed part of the contract of sale at
     auction was not incorporated into the private sale. However, the vendors did not
     attempt to argue that this variation meant that the entire sale was unauthorised.
4   [1991] 2 EGLR 9.
5   (1848) 6 Hare 443.

defendant's property was to be sold by auction. The day before the sale was due, it was discovered that the printed particulars and conditions of sale did not reserve to the vendor a right of way, as had been intended. A number of copies were therefore altered and the auctioneer read out the alteration before the bidding started. The claimant, who had an unaltered copy of the particulars and who did not hear the alteration, successfully bid for the property. The auctioneer then inadvertently drew up and signed a memorandum of the sale upon an unaltered copy of the particulars. The claimant sought specific performance of the agreement as it appeared in writing (ie with no right of way reserved), but this claim was rejected by Wigram V-C. The judge pointed out that the auctioneer's authority to sell according to the original particulars had clearly been revoked by the vendor and stated that:

> The revocation of the authority of the auctioneer is operative per se, and therefore ... is binding upon persons not parties to or conusant of it.

This decision was followed in *Re Hare and O'More's Contract*,[1] where the particulars of sale contained an inaccuracy which was orally corrected by the auctioneer. It was held that, even if the purchaser had not heard the correction (and the evidence on this point was conflicting), it would be inequitable to compel the vendor to complete the sale, since the purchaser would then have to be allowed compensation in respect of the error.

These two decisions are not, it is submitted, conclusive upon the question of the auctioneer's implied authority. Both were actions for specific performance of a contract and in each case the court, in the exercise of its equitable discretion, refused to grant a decree, a refusal which could be justified on the simple ground of hardship to the defendant. Had these purchasers sued for damages (a common law remedy) the courts' discretion would have been removed; a definite decision would then have been necessary as to whether a purchaser is entitled to assume that an auctioneer has authority to sell in accordance with the published particulars and conditions of sale.

---

1   [1901] 1 Ch 93.

Enforcement of sale by auctioneer

It is well established that, where goods are sold at auction, the auctioneer is entitled to sue in his own name to recover the price from the highest bidder.[1] This right is said to be based on the auctioneer's 'special property' in the goods, coupled with the fact that an auctioneer has a lien over the goods in order to safeguard his or her commission. However, there appears to be no equivalent right on a sale of land. In the Irish case of *Cherry* v *Anderson*[2] the claimant sold certain leasehold property for a named vendor, under a contract which made the purchaser responsible for the auctioneer's fees. An action by the auctioneer to enforce this obligation was unsuccessful, for reasons explained by Palles CB:[3]

> Upon the sale of real estate the auctioneer has no right to sue in his own name, analogous to his right on the sale of goods. In the latter case he has, or is deemed to have, a possession of the goods and a qualified property in them. On the sale of real property the right to sue depends on the written contract alone; and unless that contract be so framed as to render the auctioneer the ostensible vendor ... he cannot on the sale of real estate maintain in his own name an action against the purchaser.

The principle thus clearly enunciated had been the basis of two earlier English decisions which concerned the practice of letting land by auction. In *Evans & Thomas* v *Evans*[4] the claimant auctioneers let certain fields by auction, the defendant being the highest bidder. The conditions of sale stated that the land was let by, and the rent was payable to, the auctioneers; but the words 'Approved by me, David Jones' appeared at the bottom of the conditions. On these facts it was held that the auctioneers could not sue on the contract, which was clearly made with the real lessor, David Jones. In *Fisher* v *Marsh*[5] on the other hand, the claimant was employed by a committee responsible for organising the Oxford races to let by auction small plots of land for the erection of booths. The conditions, which were signed by the auctioneer, gave no indication that anyone else was the true lessor; accordingly the

---

1   *Williams* v *Millington* (1788) 1 H Bl 81; *Benton* v *Campbell, Parker & Co Ltd* [1925] 2 KB 410.

2   (1876) IR R 10 CL 204.

3   At p 209.

4   (1835) 3 A & E 132.

5   (1865) 6 B & S 411.

auctioneer was held entitled to maintain an action for use and occupation against a successful bidder.

Although an auctioneer is not usually likely to succeed in an action against a purchaser of land which is based on the contract of sale, there may in some circumstances be an alternative means of redress. In *Cleave* v *Moore*[1] the defendant bought certain premises at auction and signed a contract stating that he had paid a deposit. In fact no deposit was paid but the defendant gave the auctioneers an IOU for the amount in question. The auctioneers were held entitled to recover upon this IOU; the effect of the transaction, in the opinion of the Court of Exchequer, was that the deposit had been paid by means of a loan from the auctioneers to the purchaser.

A somewhat similar point arose in *Pollway Ltd* v *Abdullah*,[2] where the Court of Appeal reached a decision of great practical importance to auctioneers. In that case, which concerned the sale of certain ground rents, the conditions of sale provided for a 10 per cent deposit to be paid to the auctioneers as agents for the vendors. The defendant, who signed a memorandum of purchase as the highest bidder, made out a cheque for the auctioneers for the deposit, which they accepted. The cheque was subsequently dishonoured, whereupon the vendors rescinded the contract and resold the property. When sued upon the cheque, the defendant claimed that there had been a total failure of consideration, in that the vendors were no longer liable to complete the sale to him. The Court of Appeal, however, upheld the auctioneers' claim on the ground that they had given consideration, either by warranting their authority to sign a memorandum and accept a deposit, or by agreeing to accept a cheque in place of the cash which, strictly speaking, the defendant was bound to provide.

The bidding process

Under this heading we examine various legal principles which govern the bidding process at auctions. Where everything runs smoothly, the bidders make a series of offers of steadily increasing amounts, the last of which ripens into a contract when the auctioneer signifies acceptance by bringing down the hammer. However, this simple picture conceals a number of possibilities that things may not run smoothly, largely

---

1   (1857) 28 LT (OS) 255. See also *Hodgens* v *Keon* [1894] 2 IR 657.
2   [1974] 2 All ER 381.

because the very nature of an auction presupposes a conflict of interest. On the one hand, bidders are naturally concerned to purchase property as cheaply as they can. On the other hand, the vendor will wish to obtain as much as possible for the property and may, in addition, have a minimum price in mind below which the property cannot be sold at all.

In taking any positive steps to achieve their objectives, the bidders are hampered by the fact that they are in competition with each other, as well as with the vendor. Nevertheless, bidders can and do on occasion seek to withdraw bids, raise disputes over bids which they and other bidders have made, or claim that their bids are based on some mistake.

A vendor, by contrast, has a clearly defined interest to protect, and two long-established means of protecting it: the imposition of a reserve price, below which the property may not be sold, and the reservation of a right to bid. The second of these methods has the advantage of being more flexible, since a vendor who bids personally is in a position to assess the market and to decide at what point to cease bidding. There has never been much doubt as to the legality of these methods of self-protection for vendors, provided that their existence is brought to the attention of prospective bidders before the commencement of the sale. The problem arises when such action is taken secretly on a vendor's behalf.

Sales subject to reserve

Where it is made clear to bidders at auction that reserve prices have been (or may have been) set, the legal position is fairly straightforward. The auctioneer has no authority of any kind to sell for less than the reserve and if, through inadvertence, property is knocked down at below its reserve, the vendor is not bound by the sale. This was recognised by the Court of Appeal in *McManus* v *Fortescue*,[1] where a horse was sold at auction under conditions of sale stating that 'each lot will be offered subject to a reserve price'. The auctioneer, forgetting the reserve on this lot, knocked down the property below the reserve price, realising his mistake in time to avoid signing a memorandum of the sale. The claimant, to whom the property had been knocked down, argued that either he had a contract with the

---

1   [1907] 2 KB 1.

vendor, which the auctioneer was obliged to render enforceable by signing a memorandum,[1] or he had no contract, in which case the auctioneer was liable for breach of warranty of authority.

The Court of Appeal rejected both these claims, holding that where a sale takes place with notice to all concerned that it is subject to a reserve 'every bid, including the final one, and the acceptance of that final bid indicated by the fall of the hammer, is conditional'[2] on the reserve being reached. Thus, both the alleged contract, and the auctioneer's alleged warranty that he had authority to make it, were rendered ineffective by the failure of the condition.[3]

It appears that the principle laid down in *McManus* v *Fortescue* is equally applicable where the conditions of sale merely state that there may be reserves; this information is sufficient to put bidders on inquiry, and to render their bids conditional upon the reserve price (if any) having been reached. In *Fay* v *Miller, Wilkins & Co*[4] the defendants, a firm of auctioneers, were instructed to sell a certain house. The conditions of sale reserved to the vendor the right to fix a reserve price, and the vendor duly stipulated a figure of £750. The auctioneers, however, inadvertently knocked the property down to the claimant for £600 and signed a memorandum 'ratifying' the sale which the Court of Appeal held was sufficient to satisfy section 40 of the Law of Property Act 1925.[5] It was held that the vendor was not bound by the contract for, having received notice of the possibility of a reserve being fixed, 'the purchaser was put on inquiry, and if she chose to bid without knowing whether or not there was a reserve price, the risk was hers and she could not hold the vendor bound if in fact there was a reserve price which exceeded the amount of her bid'.[6]

In this case the disappointed purchaser was not left entirely without a legal remedy, being awarded damages against the auctioneers for breach of warranty of authority to sell. The auctioneers' argument (based on *McManus* v *Fortescue*) that everything was conditional, was rejected by the Court of Appeal, since 'a memorandum was actually signed by the auctioneers and it is impossible to treat the

---

1   Such a memorandum is no longer required: see p 157.

2   At p 7, *per* Fletcher Moulton LJ.

3   The decision on the latter point is criticised below.

4   [1941] Ch 360.

5   Since repealed by the Law of Property (Miscellaneous Provisions) Act 1989, ss 2, 4 and Schedule 2: see p 157.

6   At p 365, *per* Greene MR.

condition as applying after that was done'.[1] As Clauson LJ said:[2] 'The contract is a clear contract for a sale for £600, and, when read with the conditions, it appears to imply that the reserve, if any, is not greater than £600.'

The reasoning embodied in these two decisions of the Court of Appeal, which makes the auctioneer's liability depend upon whether or not a memorandum of sale is signed, is unfortunate, since the law no longer requires a memorandum in order to render the sale enforceable.[3] In any event, it is suggested that the approach is quite erroneous. Where a sale is expressly made subject to a reserve, an auctioneer who knocks property down is surely giving a clear assurance that the reserve has been passed and that he accordingly has authority to sell. If the reserve has not been passed, the auctioneer should be liable in damages to the highest bidder for breach of warranty of authority, irrespective of what happens thereafter.

Sales without reserve

In modern times it would be rare indeed for land or buildings to be sold at auction without a clear statement that the sale is, or at least may be, subject to a reserve price. However, a sale without reserve is not entirely unknown, and it is therefore necessary to consider two different legal problems which could emerge in such circumstances. These are, first, a purported sale by an auctioneer at less than a secret reserve price which has in fact been fixed by the vendor; and, second, an auctioneer's refusal to accept the highest bid on the ground that the property has not reached its secret reserve.

Where the auctioneer purports to sell

An important question, which has not yet been satisfactorily answered by the courts, is whether an auctioneer has implied authority to sell property 'without reserve'. If there is such authority, then a purchaser to whom a lot is inadvertently knocked down at less than the reserve price will be entitled to keep it.

---

1   At p 364, *per* Greene MR.
2   At p 367.
3   See p 157.

Some guidance on this matter may be obtained from the old case of *Bexwell* v *Christie*,[1] in which the defendant auctioneer was employed to sell 'the goods and effects of a gentleman deceased, at his house in the country, by order of the executor'. The conditions of sale stipulated 'that the goods should be sold to the best bidder'. The claimant, who had no other connection with this sale, sent a horse to the auctioneer, asking him to include it in the sale and instructing him not to let it go for less than £15. The defendant knocked down the horse for less than this sum and, when sued by the claimant for disobeying instructions, argued that to have bid on the vendor's behalf would, in the light of the conditions of sale, have been a fraud on those bidding at the sale. This argument was accepted by the Court of King's Bench, which held that the vendor's instruction, being unlawful, was not binding on the auctioneer. Lord Mansfield CJ went on to remark that the vendor might validly have achieved his object by instructing the auctioneer to put the horse up at the relevant price; since, however, he did not consider the effect in such a case of the auctioneer failing to do so, it is not clear whether such an instruction would need to be made public.

The decision in *Bexwell* v *Christie* suggests that the interests of persons bidding at an auction are to be considered when interpreting the auctioneer's instructions, and this judicial attitude is also revealed in the case of *Rainbow* v *Howkins & Sons*.[2] The defendant there was instructed to sell a pony at auction for not less than £25; forgetting this instruction, however, the defendant stated that the sale was 'without reserve' and duly knocked down the pony to the claimant for 15 guineas. Almost immediately the defendant realised this mistake and, pointing it out to the claimant and refusing to sign a memorandum of sale, he put the pony up again and bought it in for seventeen guineas. The claimant demanded delivery of the pony from the auctioneer, in accordance with the contract of sale which was made at the fall of the hammer; alternatively, the claimant sought damages for breach of the auctioneer's warranty of authority to sell at 15 guineas. A Divisional Court of King's Bench rejected the first claim, on the ground that such an action required written evidence to satisfy section 4 of the Sale of Goods Act 1893.[3] The second claim also failed because there was no

---

1   (1776) 1 Cowp 395.
2   [1904] 2 KB 322.
3   Since repealed by the Law Reform (Enforcement of Contracts) Act 1954, s 2: see p 157.

breach of warranty of authority; knocking down the pony at 15 guineas was (in the court's view) an authorised act. As Kennedy J said:[1]

> The hammer had fallen, and as between the claimant as the highest bidder and the auctioneer's principal the right of the claimant could not have been defeated by the principal shewing that the auctioneer was authorised by him to sell only subject to a reserved price. The defendant, the auctioneer, had an apparent authority, which his principal, if he had been sued by the claimant, would not have been allowed in point of law to repudiate, after a sale had been concluded by the hammer being knocked down, upon the ground that his private instructions had been contravened by the auctioneer in selling without reserve.

The decision in *Rainbow* v *Howkins & Sons* seems unexceptionable, and yet its authority is open to question, following criticism from the Court of Appeal in the later case of *McManus* v *Fortescue*.[2] In that case, which dealt with a sale which was clearly subject to reserve, Fletcher Moulton LJ stated:[3]

> A principal, therefore, who gives authority to an auctioneer to sell subject to a reserve price gives no power to the auctioneer, either expressly or impliedly, to accept a less price. The case of *Rainbow* v *Howkins*, so far as it is inconsistent with this view, cannot be regarded as in harmony with well-established principles. It is admitted in this case that the price offered by the bid of the claimant was lower than the reserve price. The auctioneer could not effectively accept such a bid, because he could not make a contract so as to bind his principal to accept less than the reserve price.

Although these remarks were quite unnecessary for the decision in *McManus* v *Fortescue*, which can easily be justified on the basis of the publication of the condition relating to reserves, it cannot be denied that they cast some doubts upon the reliance which can now be placed on *Rainbow* v *Howkins*. It is nevertheless submitted that that decision is correct; if so, since written evidence is no longer required where land is sold at public auction,[4] it follows that the fall of the hammer would be sufficient to create a binding contract between the vendor and the highest bidder.

---

1   At p 326.
2   [1907] 2 KB 1.
3   At p 6.
4   See p 157.

Where the auctioneer refuses to sell

The second potential problem arising out of a sale 'without reserve' is where the intervention of the auctioneer, as agent for the vendor, prevents the highest bidder from securing the property. The auctioneer can achieve this (and thereby give effect to a secret reserve) either by refusing to bring down the hammer at all, or by bidding for the vendor until all genuine bidders have dropped out. The question is whether, when bidders have been led to expect a sale to be without reserve, such actions on the part of the auctioneer are lawful.

Before turning to examine the English cases in which this question has been discussed, it may be of interest to note the straightforward approach which was adopted by the Scottish Court of Session in *Fenwick* v *Macdonald, Fraser & Co Ltd*.[1] In that case, where a herd of cattle was offered in lots 'for unreserved sale', the claimant bid 42 guineas for a particular bull. The auctioneer, however, intimating that there was a reserve of 150 guineas on this animal, withdrew it from the sale. The court held that the claimant could not recover damages from either the auctioneer or the seller, and Lord Trayner stated the reason for decision succinctly:[2]

> If it is the power of any competitor at an auction sale to retract his bid before the fall of the hammer, it follows that the exposer is equally entitled to withdraw his subject, because if the competitor is not bound the exposer is not bound either.

It was also held in that case that the auctioneer, who had at all times been acting upon the instructions of a named principal, incurred no personal liability to the claimant.

The Scottish position, though logical, seems somewhat questionable in moral terms, since it enables a vendor to impose a reserve price on property, while at the same time obtaining whatever benefit there may be in advertising the sale as 'without reserve'. Faced with this problem, the English courts have sought to find some way of giving effect to a 'without reserve' undertaking, while recognising that any liability must necessarily arise before a contract of sale is made; that is to say, before the hammer falls.

---

1   (1904) 6 F 850.
2   At p 854.

In the leading case of *Warlow* v *Harrison*,[1] the defendant auctioneer advertised and put up for sale a mare described as 'the property of a gentleman, without reserve, Janet Pride, a brown mare'. The claimant bid 60 guineas for this mare; there was then a bid of 61 guineas but the claimant, upon learning that this bid had been made by the seller, refused to bid further. The mare was accordingly knocked down to the seller whereupon the claimant, having tendered 60 guineas to the auctioneer and demanded the mare, sued the auctioneer for damages. The claimant's case alleged that, since he was the highest bidder, the auctioneer had become his agent for the purpose of completing the contract, and was liable for breach of this agency. The existence of the alleged agency was denied, both at first instance and on appeal, on the ground that the auctioneer's limited authority to sign a memorandum of sale as agent for the purchaser[2] arises only when the hammer falls. The defendant was therefore not liable on the case as pleaded; however, the judges took the opportunity to consider, *obiter*, the merits of the situation.

The judgment of the Court of Queen's Bench was delivered by Lord Campbell CJ, who said that, however the case was pleaded, the defendant could not be liable to the claimant upon any contract, since this would contradict the well-established principle of *Payne* v *Cave*.[3] When the claimant appealed, however, the five judges of the Exchequer Chamber all thought that an auctioneer could be liable in such circumstances, although they were divided in their reasoning. In giving the decision of the majority, Martin B likened the situation to a 'reward case', in which a general offer ripens into a contract with anyone who performs the stated conditions. He continued:[4]

> Upon the same principle, it seems to us that the highest bona fide bidder at an auction may sue the auctioneer as upon a contract that the sale shall be without reserve. We think the auctioneer who puts the property up for sale upon such a condition pledges himself that the sale shall be without reserve; or, in other words, contracts that it shall be so; and that this contract is made with the highest bona fide bidder; and, in case of a breach of it, that he has a right of action against the auctioneer ... We think the auctioneer has contracted that the sale shall be without reserve; and that

---

1   (1859) 1 E & E 295.
2   Such authority is no longer of practical significance, since written evidence is no longer required to render a sale of either land or goods enforceable: see p 157.
3   (1789) 3 TR 148: see p 193.
4   (1859) 1 E & E 295, 316.

the contract is broken upon a bid being made by or on behalf of the owner, whether it be during the time when the property is under the hammer, or it be the last bid upon which the article is knocked down; in either case the sale is not 'without reserve', and the contract of the auctioneer is broken.

The two minority judges, Willes and Bramwell JJ, preferred to rest their decision as to the auctioneer's potential liability upon the ground of breach of warranty of authority, that is to say, upon the auctioneer's untrue claim that he was authorised by the vendor to sell 'without reserve'.

Before considering the merits of these two lines of reasoning, one point should be noted: no action was brought against the seller in this case and the courts made no mention of this possibility, preferring to treat any liability which might arise as attaching to the auctioneer. It would seem, however, as a matter of principle, that a vendor who causes the conditions of sale to be violated, either by making an unlawful bid or by instructing the auctioneer to withdraw property from the sale, should be equally liable to the disappointed highest bidder. In any event, it was made clear that the legal responsibility would ultimately rest with the seller. As Martin B said:[1]

> We entertain no doubt that the owner may, at any time before the contract is legally complete, interfere and revoke the auctioneer's authority: but he does so at his peril; and if the auctioneer has contracted any liability in consequence of his employment and the subsequent revocation or conduct of the owner, he is entitled to be indemnified.

The minority view in *Warlow v Harrison*, that the auctioneer is liable for breach of warranty of authority, might seem superficially attractive, but the case does not really fall within the normally accepted limits of that principle. Breach of warranty of authority typically arises where an agent purports to make a contract on behalf of a client but is later found to have acted without authority. In such circumstances the alleged contract is not binding upon the agent's principal, and the other party is entitled to sue the agent personally for breach of an implied representation of authority.[2] In the *Warlow v Harrison* situation, the highest bidder's complaint is not that the auctioneer has done something without authority, but that the auctioneer has refused to exceed his or her authority.

---

1   At p 317.
2   See, for example, *Fay v Miller, Wilkins & Co* [1941] Ch 360: p 175.

It is therefore apparent that, if the disappointed bidder is to have a right of action, this must depend upon some form of contractual obligation, which is created by making a bid on the faith of the statement that the sale will be 'without reserve'.[1] This is the principle which was laid down by the majority in *Warlow* v *Harrison*, and it has been the subject of judicial consideration in a number of later cases.

In *Mainprice* v *Westley*,[2] the defendant auctioneer advertised that a shop was to be 'offered for peremptory sale by auction, by direction of the mortgagee with a power of sale'. The advertisement named the mortgagee's solicitor as the person from whom further particulars could be obtained. The claimant bid for the shop but, after bidding had started, it was announced that there was a reserve price on the property, and the mortgagee's solicitor bought it in at this price. The claimant brought an action against the auctioneer on the basis of *Warlow* v *Harrison*, but this claim failed. The court ruled that, in contrast to what had occurred in *Warlow* v *Harrison*, the auctioneer here had identified his client and could not therefore be said to have made any implied contract with the claimant.

It is suggested, for two reasons, that this alleged ground of distinction between the cases is not a valid one. First, the principal in *Warlow* v *Harrison* was not undisclosed, but merely unnamed, and it is well established that an agent does not automatically incur personal liability merely by failing to name his principal. Second, two of the three judges in *Mainprice* v *Westley* itself expressly doubted whether an auctioneer could ever be said to act for an undisclosed principal:

> Inasmuch as the character of an auctioneer as agent is unlike that of many other agents, as to whom, so long as the fact of there being a principal is undisclosed, it remains uncertain whether the contracting party is acting as principal or agent; while in the employment and duty of an auctioneer the character of agent is necessarily implied, and the party bidding at the auction knowingly deals with him as such, and with the knowledge that his authority may be at any moment put an end to by the principal.[3]

---

1   It is not of course contended that the contract of sale is created then; see the Sale of Goods Act 1979, s 57(2). In *British Car Auctions Ltd* v *Wright* [1972] 3 All ER 462 it was specifically held, for the purposes of criminal law, that an auctioneer does not 'offer' goods for sale.

2   (1865) 6 B & S 420.

3   (1865) 6 B & S 420, 429, *per* Blackburn J.

The majority decision in *Warlow* v *Harrison* received a degree of support in two later English cases. In *Harris* v *Nickerson*,[1] the defendant auctioneer advertised that certain brewing materials, plant and office furniture would be sold by him on a certain date. The claimant obtained a commission to purchase the furniture and travelled from London to Bury St Edmunds, only to find that the furniture had been withdrawn from the sale. In rejecting the claimant's action for loss of time and expenses, the Court of Queen's Bench distinguished *Warlow* v *Harrison* on the ground that the present advertisement was not an offer, but merely a declaration of intention giving rise to no legal rights on the part of anyone who relied upon it. However, both Blackburn J and Quain J thought that a contract might arise between an auctioneer and the highest bidder in the circumstances of *Warlow* v *Harrison*, although Blackburn J appeared to prefer the argument based on breach of warranty of authority.

In *Johnston* v *Boyes*,[2] an action was brought against a vendor of land for refusing to allow the highest bidder, to whom it had been knocked down, to sign a contract of purchase and to pay a deposit by cheque. If this refusal could not be justified[3] it would clearly amount to a breach of the conditions of sale, which provided that the highest bidder should be the purchaser. In dealing with this possibility, Cozens-Hardy J said:[4]

> A vendor who offers property for sale by auction on the terms of printed conditions can be made liable to a member of the public who accepts the offer if those conditions be violated: see *Warlow* v *Harrison*.

The 19th-century authorities cited above were carefully reviewed by the Court of Appeal in *Barry* v *Heathcote Ball & Co (Commercial Auctions) Ltd*,[5] where the defendant auctioneers sold goods on behalf of the Customs and Excise in satisfaction of a tax debt. The goods included two new engine analysers, with a list price of £14,000 each, which were put up for sale 'without reserve'. The only bid for the machines was

---

1   (1873) LR 8 QB 286.
2   [1899] 2 Ch 73.
3   In fact it was; the auctioneers, who recognised the bidder as someone without substantial resources, were held entitled to insist on payment of the deposit in cash.
4   [1899] 2 Ch 73, 77.
5   [2001] 1 All ER 944.

one of £200 each from the claimant, but the auctioneer refused to sell at such a low price. The Court of Appeal, approving the reasoning in *Warlow* v *Harrison* and rejecting the criticism of it in *Mainprice* v *Westley*, held the auctioneers liable for breach of a collateral contract with the claimant, based on the bidder having given consideration by attending the auction and making a bid. The damages for this breach, reflecting the difference between the bid and the court's view of market value of the goods, were assessed at £27,600.

Bids by or on behalf of the vendor

Reservation of the right to bid

From an early stage, common law insisted that a vendor was only permitted to bid where the right to do so was explicitly notified. Where a vendor had not expressly reserved the right to bid personally or through an agent, any such bid would constitute a fraud upon those bidding at the sale. In *Bexwell* v *Christie*[1] for example, a sale of goods was held under conditions of sale which stated 'that the goods should be sold to the best bidder'. The claimant sent his horse for inclusion in this sale, with an instruction to the defendant auctioneer that he should not let it go below a certain price. The auctioneer disobeyed this instruction, and the Court of King's Bench held that he was justified in so doing, for to have bought in the horse in these circumstances would have been a fraud upon the sale.

In accordance with this principle, a person who was induced to pay an inflated price at an auction by bids which, though apparently independent, were actually made on behalf of the vendor, was not bound by the sale and could recover any deposit paid.[2] In *Green* v *Baverstock*[3] the defendant bought goods at an auction, a condition of which was that the highest bidder should be the purchaser. Upon later discovering that a puffer employed by the seller had bid against him, the defendant refused to remove or pay for the goods and the auctioneers failed in their action against him. Byles J stated the common law principle:[4]

---

1   (1776) 1 Cowp 395. See also *Narramore* v *Fuller, Hall & Foulsham* (1932) 76 Sol Jo 289.
2   *Thornett* v *Haines* (1846) 15 M&W 367.
3   (1863) 14 CBNS 204. See also *Howard* v *Castle* (1796) 6 TR 642; *Crowder* v *Austin* (1826) 3 Bing 368; *Wheeler* v *Collier* (1827) 1 Mood & M 123;   *R* v *Marsh* (1829) 3 Y & J 331.
4   At p 208.

> Upon a sale by auction where the highest bidder is to be the purchaser, the secret employment of a puffer by the vendor is a fraudulent act. The sale is vitiated by the fraud, and void.

The position of common law may have been clear, but that of equity was less so. The courts here, while equally anxious to suppress fraud, were also concerned to prevent the sale of property at a gross undervalue. Accordingly, the rule developed in equity that, even without disclosing the fact, a vendor might validly appoint one person to bid, provided that this was not otherwise inequitable.[1] This rule, however, did not apply where the sale was expressly stated to be 'without reserve';[2] nor did it permit the secret appointment of more than one puffer, since the effect of this would clearly be to enhance the price rather than merely to prevent a sale at an undervalue.[3]

The conflict between law and equity, although limited in scope, was none the less undesirable and its resolution, at least as far as the sale of land was concerned, was one of the reasons for the enactment of the Sale of Land by Auction Act 1867. The Act gave priority to the rule of common law, which was generally accepted as being preferable.[4]

Section 4 of the Act provides:

> And whereas there is at present a conflict between Her Majesty's Courts of law and equity in respect of the validity of sales by auction of land where a puffer has bid, although no right of bidding on behalf of the owner was reserved, the courts of law holding that all such sales are absolutely illegal, and the courts of equity under some circumstances giving effect to them but even in courts of equity the rule is unsettled: and whereas it is expedient that an end should be put to such conflicting and unsettled opinions: Be it therefore enacted that from and after passing of this Act whenever a sale by auction of land would be invalid at law by reason of the employment of a puffer, the same shall be deemed invalid in equity as well as at law.

Had the Act stopped at this point, the position would simply have been that the rules of common law, which were reasonably clear,

---

1   *Conolly* v *Parsons* (1797) 3 Ves 625n; *Bramley* v *Alt* (1798) 3 Ves 620; *Smith* v *Clarke* (1806) 12 Ves 477; *Woodward* v *Miller* (1845) 2 Coll 279.

2   *Meadows* v *Tanner* (1820) 5 Madd 34.

3   *Mortimer* v *Bell* (1865) 1 Ch App 10.

4   Even by some Chancery judges: see the remarks of Lord Cranworth LC in *Mortimer* v *Bell* (1865) 1 Ch App 10, 16.

prevailed over those of equity, which were not. As Lindley J said in *Parfitt* v *Jepson*:[1] 'With regard to the statute, it seems to me that the object of the 4th section is not to restrict the legal doctrine, but to leave it as it was and to make the equitable doctrine correspond with it.' Unfortunately, however, the opportunity was taken to restate the common law rules and the results, contained in sections 5 and 6, were so clumsily drafted[2] that they have merely increased the confusion which they were intended to prevent.

Section 5 provides:

> And whereas as sales of land by auction are now conducted many of such sales are illegal, and could not be enforced against an unwilling purchaser, and it is expedient for the safety of both seller and purchaser that such sales should be so conducted as to be binding on both parties: Be it therefore enacted by the authority aforesaid as follows: that the particulars or conditions of sale by auction of any land shall state whether such land will be sold without reserve, or subject to a reserved price, or whether a right to bid is reserved; if it is stated that such land will be sold without reserve, or to that effect, then it shall not be lawful for the seller to employ any person to bid at such sale, or for the auctioneer to take knowingly any bidding from any such person.

Section 6 provides:

> And where any sale by auction of land is declared either in the particulars or conditions of such sale to be subject to a right for the seller to bid, it shall be lawful for the seller or any one person on his behalf to bid at such auction in such manner as he may think proper.

In requiring any reserve price or bidding right to be stipulated in the particulars or conditions of sale, the Act has clearly altered the common law, which allowed this information to be given to bidders by other means, most commonly by an announcement from the auctioneer. However, the Act does not mention any sanction for failure to furnish this information in the stipulated manner. It is submitted that, since the purpose of the Act was to clarify the law relating to puffers, rather than to amend conveyancing techniques, such default would not render the sale itself invalid; what it would mean is that the

---

1   (1877) 46 LJCP 529 533.
2   The Act is hardly a model piece of drafting; section 3 provides a definition of 'agent', a word which makes no other appearance in the Act!

sale would then be treated as being 'without reserve', so as to preclude any bidding or other interference on behalf of the vendor.

Exactly what must be included in the particulars or conditions of sale, in order to satisfy section 5 of the Act, has been a matter of some controversy as yet unsettled by the courts. It is not clear whether a vendor who wishes to reserve a right to bid must, in addition to making express provision for such a right, also state that the sale is subject to a reserve, or whether the reservation of a right to bid automatically carries that implication. Support for the former view[1] is to be found in *Gilliat* v *Gilliat*,[2] the first case to be decided upon the Act. In that case property was put up for sale under a decree of the court. The conditions of sale state: 'The sale is subject to a reserved bidding which has been fixed by the judge.' The property was knocked down to a purchaser for £29,000, the amount of the reserved bidding, after a person employed by the auctioneer had bid up to £28,900. Although, in the absence of the puffer, the purchaser could not have obtained the property at a lower price, the sale was none the less set aside. Lord Romilly MR said:[3]

> I think the Act makes a distinction between a reserved bidding and a reserved right to bid. It say that you must state whether there is a reserved price or not, and further, if you state that there is a reserved price, you must also state that a right to bid is reserved in order that you may employ a person to bid on your behalf.

The second possible interpretation of section 5 is that, since it lays down three possibilities, each one separated by the word 'or', it is sufficient if any one of these is mentioned in the particulars or conditions of sale.[4] In this connection some assistance may be gained from the case of *Dimmock* v *Hallett*,[5] although this is not of direct authority since it was decided prior to the Act. In that case, which concerned the sale by a mortgagee of the mortgaged property, the auctioneer announced that the sale was without reserve, but that the parties interested in the estate were at liberty to bid. The defendant bought the property for £19,000 but refused to complete the sale on

---

See D Macintyre, *Law Relating to Auctioneers and Estate Agents*, pp 64–6.

[2] (1869) LR 9 Eq 60.

At p 62.

See *Bateman's Law of Auctions*, 11th ed, pp 143-8.

(1866) 2 Ch App 21.

the ground that, from £14,000, the only other bidder had been the mortgagee himself. It was held that the purchaser could not avoid the sale on this ground. Turner LJ said:[1]

> The question then remains, what meaning is to be attributed to the statement that a sale is without reserve, but that the parties interested are at liberty to bid. The two branches of the statement are not very consistent, but I think that they may be read together by taking the second as a qualification of the first; and if a purchaser knows that parties interested have liberty to bid, he cannot be entitled to be discharged on the ground that they have bid against him.

It is submitted that the formal requirements of the Act are to be interpreted in the light of its intention, which was clearly to prevent purchasers of land at auction from being defrauded. This object is satisfactorily achieved, as *Dimmock* v *Hallett* clearly shows, by an express reservation of a right to bid, either with or without a further statement that the sale is subject to a reserved price. The view of Lord Romilly MR[2] to the contrary is, it is submitted, no more than a passing remark since, in the case before him, no right to bid had been reserved, and the employment of a puffer was therefore unlawful.

Whether or not it is strictly necessary, the prudent auctioneer will ensure that, wherever a right to bid is reserved, the particulars or conditions of sale also contain an express statement to the effect that the sale is subject to a reserve. It has been suggested that the reservation of the right to fix a reserve price would not be sufficient, as this would not inform bidders whether such a reserve had in fact been fixed;[3] however, in other circumstances such a condition has been held sufficient to put a purchaser on inquiry,[4] and it is submitted that a similar conclusion would be reached in this context.

The Sale of Land by Auction Act 1867 has come before the courts on two subsequent occasions. In *Parfitt* v *Jepson*,[5] the lease of a brickyard was auctioned under conditions which stipulated that the highest bidder should be the purchaser, but that the vendor or his agent should have the right of bidding once for the property. The auctioneer

---

1   At p 26.
2   *Gilliat* v *Gilliat* (1869) LR 9 Eq 60, 62: p 187.
3   Macintyre, *op cit*, p 63.
4   See *Fay* v *Miller, Wilkins & Co* [1941] Ch 360: p 175.
5   (1877) 46 LJCP 529.

bid three times in competition with a third party, the last bid being £320, and then asked the vendor, who was present at the sale, to state his reserve price. The vendor said that his reserve was £350; the defendant, at the auctioneer's request, bid £351, whereupon the property was knocked down to him. The Court of Common Pleas held that the defendant was entitled to refuse to complete the sale, both on common law principles (as preserved by section 4 of the Act) and in accordance with the provisions of section 5 and 6. As far as the common law was concerned, it was held that the bids made by the auctioneer were to be treated as made by a puffer. Furthermore, the court gave short shrift to the argument that the effect of these unlawful bids had been wiped out by the subsequent statement of the reserve price:[1]

> It is impossible to say to what extent a purchaser may not be influenced by bids, whether made earlier or not. People at a public auction are often influenced by the amount of competition going on, and where they assume the bids to be bona fide they are often willing to give more for the property than they would if they knew them to be sham bids.

With regard to the statutory provisions, it was held that, although the reservation of a right to bid once did not come within the express words, it was within the spirit and meaning of sections 5 and 6. In consequence, any action in excess of the rights which were expressly reserved was void. Moreover, Grove J was of the opinion that a right to bid once was very like a reserve price, something which does not of itself entitle the seller to bid, either personally or through an agent.

It may also be noted that, although the issue did not arise for the decision in the case itself, Grove J also stated his opinion that the Act had abolished altogether the practice of employing more than one puffer, since this would create a false appearance of competition.

The only other case concerning these statutory provisions is *Hills & Grant, Ltd v Hodson*,[2] in which Luxmoore J had to consider what precise form of words is required to satisfy section 5 of the Act. In that case freehold building land was being sold in three lots, of which the present action concerned Lot 2. The Special Conditions of Sale incorporated by reference the National Conditions of Sale (11th edition); these provided that, subject to a right of the vendor to bid up to a reserved price, the highest bidder should be the purchaser. The

---

1   *Per* Lindley J, at p 533; see also Grove J at p 532.
2   [1934] Ch 53.

Special Conditions further provided that, in the event of Lot 2 remaining unsold or being bought in by the vendors, the vendors should be entitled to withdraw Lot 3 from the sale. The defendant, to whom Lot 2 was knocked down, refused to complete the sale on the ground that a puffer had bid against him, claiming that this was illegal in the absence of a clear statement that the property was subject to a reserved price. In dealing with this contention, Luxmoore J said:[1]

> In my judgment the Act does not require that the words 'with reserve' or 'without reserve', and no others shall be used in the particulars or conditions of sale ... the conditions of the Act are complied with so long as it is made plain, by whatever words may be chosen, that the sale is subject to a reserve.

When the conditions of sale were looked at, it was quite clear that a reserve had been fixed, as only in that event could the property possibly be 'bought in'. Thus the employment of a puffer was valid and the defendant was bound by his purchase.

To summarise, the effect of the Sale of Land by Auction Act 1867 appears to be as follows.

- A puffer (including the auctioneer) may not bid on behalf of the vendor unless a right to bid is expressly reserved in the particulars or conditions of sale.

- A similar restriction will be imposed upon a vendor who wishes to bid personally.[2] However, it may be noted that, while section 6 of the 1867 Act makes it lawful for 'the seller or any one person on his behalf' to bid where a right to do so is reserved in the appropriate manner, section 5 merely provides that, where no such right is reserved, the seller may not 'employ any person to bid'. It is thus possible to argue, albeit faintly, that personal bidding is not ruled out.

- As to the possibility that the Act also requires a clear statement (in whatever form) that a reserve has been fixed, this question remains to be decided in some future case.

---

1   At p 61.
2   *Dimmock* v *Hallett* (1866) 2 Ch App 21 proceeded on this assumption.

Fictitious bids

What now falls to be considered is the legal position where an auctioneer, whether or not in the purported exercise of a right to bid as the vendor's agent, calls bids which have not really been made at all. In *Heatley* v *Newton*,[1] leasehold property was sold at auction under conditions of sale which stated that, subject to a right of the vendor or his agent 'to bid once or oftener', the highest bidder should be the purchaser. After an apparently brisk and eager competition the claimant, who had made no previous bids, offered £13,000 and the property was knocked down to him at that price. The claimant subsequently discovered that all bids above £8,000 had been made on behalf of the vendor, and that a number were entirely fictitious, having been taken 'off the wall' by the auctioneer. On realising this, the claimant sued the seller and the auctioneer for the return of his deposit with interest. The case as reported did not deal in any detail with the merits of the claim, being mainly concerned with whether the auctioneer was entitled to pay the deposit into court and have proceedings against him stayed. Nevertheless the Court of Appeal was clearly of the opinion that the auctioneer's conduct was wrongful and that, assuming the alleged facts to be true, an action would lie against him.[2]

Notwithstanding *Heatley* v *Newton*, it is sometimes asserted by practising auctioneers that it is perfectly legitimate to 'trot' the bidding, or to take bids 'off the wall', provided only that the property has not reached its reserve. It is submitted that this is a dangerous fallacy, for at least three reasons.[3] First, there can be no doubt that inducing someone to buy property by pretending that others have made offers for it is a misrepresentation; if this is done knowingly, it is fraud.[4] Second, as we have already seen, unlawful conduct by vendor or auctioneer is no less unlawful where it takes place below the reserve, because it may continue to influence bidding above this figure.[5] Third, provided the necessary element of 'dishonesty' can be established, there seems no reason to doubt that 'calling' a bid which

---

1   (1881) 19 ChD 326.
2   Presumably the employment of more than one puffer would have been another ground for invalidating the sale, although this was not expressly mentioned.
3   Of course, the difficulties involved in proving such allegations may be considerable.
4   *Mullens* v *Miller* (1882) 22 ChD 194.
5   *Parfitt* v *Jepson* (1877) 46 LJCP 529: p 188.

the auctioneer knows has not been made will constitute the criminal offence of obtaining (or attempting to obtain) property by deception.[1]

The principle which underlies all the cases mentioned above is that the purchaser is to be protected against any unfair interference with the bidding. To this end, relief has been given where the interference in question did not take the form of puffing as such, but where the court felt that the overall result was much the same. In *Robinson* v *Wall*[2] Lord Mostyn wished to purchase certain connected estates, one of which was to be auctioned 'without reserve'. It was accordingly agreed with the vendors that Lord Mostyn should attend the sale and bid for the property; that he should pay £35,000 for it even if it were knocked down for less; and that, if the price exceeded this figure, Lord Mostyn should be free to bid higher or not, as he wished. The defendant, to whom the property was knocked down for £49,800, discovered this arrangement and refused to complete the purchase. It was held that the defendant was justified in his refusal; as Lord Langdale MR said,[3] the auction 'was, from the beginning, tainted with reserve, whilst it professed to be "without reserve".' Lord Cottenham LC, in affirming this decision, said:[4]

> Now, what took place in this case, although the arrangement is rather complicated, appears to me to amount to a reserve, or at least to precisely the same thing, so far as the public is concerned, which is the only way you can look at it. For it is quite immaterial what are the precise terms of the arrangement between the vendor and any other person, that being only the machinery by which the effect is produced. We must look to see what is the effect of what took place as regards the public — as regards those who attended the sale.

The rule preventing interference with the bidding is thus a strict one, in that it does not apply only where the vendor intends to defraud bidders at the sale. However, it appears that the purchaser can only avoid the sale where the vendor is proved to have known of the bids in question. Thus in *Parfitt* v *Jepson*,[5] counsel for the defendant

---

1   Theft Act 1968, s 15. The maximum penalties are a fine (unlimited in amount) or imprisonment for up to 10 years.
2   (1846) 10 Beav 61; affirmed (1847) 2 Ph 372.
3   (1864) 10 Beav 61, 72.
4   (1874) 2 Ph 372, 376.
5   (1877) 46 LJCP 529.

contended that the authority of a puffer to bid for the vendor must be expressly proved,[1] and this was accepted by both Grove J and Lindley J. The point arose specifically in the case of *Union Bank* v *Munster*,[2] where mortgaged property was sold at auction by its equitable mortgagees. The defendant, having already bid more than the reserved price, was forced considerably higher by a person who, with no intention of completing a purchase, was bidding on behalf of the mortgagor. The defendant claimed to avoid the sale on the ground of fraud, but the mortgagees were granted a decree of specific performance was awarded against him. As Kekewich J said:[3]

> For a man to go into an auction room and bid either a large or a small sum for a property when he knows that it is impossible for him to perform the contract is, of course, dishonest. But there is no question here of any privity of the vendors either in person or through the auctioneer or otherwise, with that dishonesty.

Thus the independent fraud of a third party, even where it causes hardship to a genuine bidder, is not the responsibility of the vendor. The latter's duty is merely to see, as far as possible, that the auction is properly and honestly conducted.

Withdrawal of bids

The conventional analysis of a sale by auction is that it consists of a series of offers from bidders, culminating in a contract when the highest bid is accepted by the auctioneer. This would suggest that, until the hammer falls, any bid may validly be withdrawn. Such reasoning is indeed supported by the old case of *Payne* v *Cave*,[4] where the defendant bid 40 shillings for a worm-tub and a pewter worm. When the auctioneer 'dwelt on the bidding', claiming that the worm was worth much more, the defendant asked the auctioneer to warrant its weight and, upon the auctioneer's refusal to do so, withdrew his bid. The defendant purchased the goods for 30 shillings at a later sale, whereupon the seller sued him for the difference in price. Rejecting the

---

1 Relying on *Thornett* v *Haines* (1846) 15 M & W 367, which proceeded on this assumption.
2 (1887) 37 ChD 51.
3 At p 53.
4 (1789) 3 TR 148.

claimant's argument that each bidder becomes a conditional purchaser, subject to the possibility of being overbid, the Court of King's Bench said:[1]

> The auctioneer is the agent of the vendor, and the assent of both parties is necessary to make the contract binding; that is signified on the part of the seller by knocking down the hammer, which was not done here till the defendant had retracted. An auction is not unaptly called 'locus poenitentiae'. Every bidding is nothing more than an offer on one side, which is not binding on either side till it is assented to. But according to what is now contended for, one party would be bound by the offer, and the other not, which can never be allowed.

The actual decision in *Payne* v *Cave*, that a bid may validly be retracted before it has been accepted, has been given statutory force by the Sale of Goods Act 1979, section 57(2). While there is no equivalent statutory provision applicable to sales of land, it is not suggested that they are to be treated differently. None the less, the basic principle cannot be regarded as entirely free from qualification. What, for example, is the effect (if any) of the frequently encountered condition of sale which states that no bid may be retracted? The conventional answer is that such a condition is meaningless, since it only becomes enforceable when it is contained in a contract, and if a bid is withdrawn before the hammer falls then there is no contract. However, there is some attraction in the argument that, if the vendor is legally obliged to adhere to the published conditions of sale,[2] then bidders should be bound as well. Nevertheless, there is little support in the case law for this argument, except perhaps for the rather special situation which came before the court in *Freer* v *Rimner*.[3] Property there was sold by order of the court, and the mortgagee of the property, although not a party to the law-suit in question, gave his consent to the sale. At the auction, a condition of which was that no bid could be withdrawn, a bid was made (and then retracted) by the mortgagee's solicitor. It was held that, having bid in full knowledge of the conditions, the solicitor was bound to complete the purchase.

---

1   At p 149.
2   See p 181.
3   (1844) 14 Sim 391.

Mistaken bids

The reluctance of English law to release a party from contractual obligations on the ground of some mistake is well known. At common law, a contract which is entered into on the basis of some fundamental mistake can be declared void, but a mistake of sufficient significance is very difficult to establish. Equity takes a somewhat more relaxed view over this issue, but the relief which equity can offer is in most cases less complete than that of common law.

The basic position outlined above is as applicable to sales by auction as to contracts formed in any other way. As Baggallay LJ pointed out in the auction case of *Tamplin* v *James*:[1]

> Where there has been no misrepresentation, and where there is no ambiguity in the terms of the contract, the defendant cannot be allowed to evade the performance of it by the simple statement that he has made a mistake. Were such to be the law the performance of a contract could rarely be enforced upon an unwilling party who was also unscrupulous.

When the Court of Appeal affirmed Baggallay LJ's decision, James LJ said:[2]

> For the most part the cases where a defendant has escaped on the ground of a mistake not contributed to by the claimant, have been cases where a hardship amounting to injustice would have been inflicted upon him by holding him to his bargain, and it was unreasonable to hold him to it.

The practical result of such judicial attitudes may be seen in *Van Praagh* v *Everidge*,[3] where the defendant, who had wished to purchase Lot 2 at a sale of land, bid by mistake for Lot 1. Kekewich J, having reviewed the relevant case law, held that there was no reason to relieve the defendant from his contract and awarded specific performance to the vendor.

Notwithstanding the general principle, there are undoubtedly circumstances in which a bidder may obtain relief from the consequences of a mistaken bid. It seems in particular that, where a bidder's mistake is in some way induced by the vendor or the auctioneer, for example through misleading sale documents, specific

---

1   (1879) 15 ChD 215, 217.
2   (1880) 15 ChD 215, 221.
3   [1902] 2 Ch 266; reversed on other grounds [1903] 1 Ch 434. Common law is even less likely to grant relief: see *Robinson, Fisher & Harding* v *Behar* [1927] 1 KB 513.

performance of the resulting contract will not be decreed.[1] Nor will equity permit a person to snap up an offer which he or she must have realised was not intended, even though that person may not have been responsible for the mistake. Thus in *Webster* v *Cecil*[2] the defendant, who had already refused an offer of £2000 for a property from the claimant, wrote to the claimant offering to sell it to him for £1,250! When the claimant purported to accept this offer, notwithstanding the obvious mistake which it contained, a decree of specific performance was refused.

In *Hartog* v *Colin & Shields*,[3] where a somewhat similar attempt was made to snap up an unintended offer, a judge went so far as to hold the resulting 'contract' void at common law.[4] However, such an extreme result is normally reserved for cases where the effect of a mistake is so fundamental that there is in truth no agreement at all between the parties.[5] In *Scriven Bros* v *Hindley*,[6] for example, bales of hemp and bales of tow were put up for auction on behalf of the claimants. Each set of bales bore the same shipping mark, for the entire consignment had arrived on the same ship (a unique occurrence, according to the expert evidence) and the defendants' buyer, who had not inspected the samples at the sale room, thought that all the bales contained hemp, which was a much more valuable commodity. The catalogue merely referred to the bales by numbers so that, when the auctioneer put up the tow, the buyer made an extravagant bid for it. The defendants were held not liable to accept the goods or pay the price, for, as A T Lawrence J said:[7]

> The parties were never *ad idem as* to the subject matter of the proposed sale; there was therefore in fact no contract of bargain and sale.

Very occasionally, a mistaken bid may qualify for some relief, even though the fault is entirely that of the bidder. In *Malins* v *Freeman*,[8] for

---

1   See, for example, *Manser* v *Back* (1848) 6 Hare 443.
2   (1861) 30 Beav 62.
3   [1939] 3 All ER 566.
4   The duty of an auctioneer in relation to surprising bids was considered by the Court of Appeal in *Friedrich* v *A Monnickendam Ltd* (1973) 228 EG 1311.
5   The classic case is *Raffles* v *Wichelhaus* (1864) 2 H & C 906, where the parties were irreconcilably at cross-purposes.
6   [1913] 3 KB 564.
7   At p 568.
8   (1837) 2 Keen 25.

example, where the defendant was employed to bid up to the reserved price on a particular lot, he bid for a different lot by mistake. Notwithstanding that the defendant alone was to blame for this error, a decree of specific performance was refused, and the vendor was left to pursue his common law claim for damages.

Bidding disputes

As mentioned previously, the essence of an auction is that a contract of sale is created when the auctioneer signifies acceptance of the last of a series of bids. It follows that, once the hammer falls, both vendor and purchaser are in principle bound by their contract, and that the auctioneer has no power to reopen matters, for example because a hitherto unnoticed bidder would like to enter the contest.

However, in the hubbub of a crowded saleroom, things are not always so clear-cut. The task of identifying who is actually bidding at any given moment is an art in itself and, not surprisingly, the fall of the hammer is sometimes closely followed by an argument. It may be, for example, that the person identified by the auctioneer as the highest bidder claims not to have made the bid in question; or that someone else claims to have made a bid which the auctioneer has ignored. In order to avoid such problems it is sensible to provide, and auction conditions commonly do provide, for any bidding dispute to be settled at the absolute discretion of the auctioneer.

The efficacy of such a condition fell to be tested in the case of *Richards* v *Phillips*,[1] which arose out of what a member of the Court of Appeal described as 'a comedy of errors of a remarkable kind'. In that case, which concerned the sale of a London theatre at the London Auction Mart, the auctioneer bid close to the reserve price on behalf of the vendors, whereupon two genuine bidders simultaneously bid above the reserve. The auctioneer saw only one of these bids, that of the claimant, and duly knocked the property down to him; however, the other bidder, a Mr Drummy (who had thought that it was his bid which had been accepted), then came forward to protest. The auctioneer, on hearing from members of his staff that Mr Drummy had indeed made a bid, decided to put the property up for sale again, relying on a condition of sale which stated that 'if any dispute arises

---

1   [1969] 1 Ch 39.

respecting a bid, the auctioneer may determine the dispute, or the property may, at the vendor's option, either be put up again at the last undisputed bid or be withdrawn'. The two rivals then bid against each other until the claimant was again successful, but this time at a price almost 50 per cent higher than on the first occasion.

The claimant duly sued the auctioneer for the difference between the two hammer prices, asserting that the auctioneer owed him a duty to provide an enforceable contract. When the auctioneer sought in defence to rely on the condition of sale quoted above, the claimant argued that, since Mr Drummy had never succeeded in attracting the auctioneer's attention, he had not in fact made any bid and there was thus no 'dispute'. This argument failed to convince the Court of Appeal, which held that this was precisely the kind of situation which the auction condition was intended to cover, and that the auctioneer was accordingly entitled to deal with the matter as he had.

Misconduct by bidders

In considering the bidding process at auction, we have so far been concerned mainly with conduct of the vendor (or the auctioneer acting on the vendor's behalf) which in some way operates to the disadvantage of a bidder. However, there are ways in which bidders may themselves seek to gain the upper hand, and some of these are clearly illegal. Three in particular merit attention.

False statements

An unscrupulous bidder, by misrepresenting the state of property which is put up for sale, may dissuade others from bidding and thereby have the property knocked down at an advantageous price. In such circumstances, the successful bidder will not be entitled to specific performance against the vendor.[1] Indeed, there seems no reason to doubt that the vendor in such circumstances will be entitled to whatever remedies are appropriate for misrepresentation, both at common law and under the Misrepresentation Act 1967.

---

1   *Howard* v *Hopkyns* (1742) 2 AtK 371.

In cases where, as a result of a bidder's denigration of property, it fails to sell altogether, an action for misrepresentation will not lie.[1] However, assuming that the statement has been made fraudulently, as it usually will have been, the person making it is guilty of the tort of malicious falsehood, more commonly (though less accurately) known as slander of title. In *Mayer* v *Pluck*[2] the defendant, an estate agent who had made previous unsuccessful offers for the claimant's house, attended the auction at which the property was to be sold and asked the auctioneer publicly whether he was aware that the house was built on an underground stream and had six inches of water in the cellar. Not surprisingly, no bids were forthcoming and the claimant was forced to remain in the house and to lose money by calling off the purchase of a new house, negotiations for which had reached an advanced stage. Milmo J told the jury that, in order to find in favour of the claimant, they must be satisfied on four points: that the defendant made the statements; that they were untrue; that they were made maliciously, with a dishonest or improper motive; and that the claimant had suffered financial loss as a result. Upon the jury finding all these matters proved, the claimant was awarded damages.

Damping the sale

Even where there has been no misrepresentation as such, the fact that property has been knocked down at an artificially low price may persuade a court of equity, in the exercise of its discretion, to refuse a decree of specific performance of the sale. In *Twining* v *Morrice*,[3] for example, the defendants, as executors of a deceased person's estate, put some land up for sale, having agreed to a reserve of £2,000 on the first lot. The defendants' solicitor knew of the reserve, although he was not instructed to bid; consequently, when the claimant asked the solicitor to bid for him, the latter bid the amount of the reserve and the property was knocked down. Evidence was given that other persons at the sale refrained from bidding against the solicitor since they believed him to be a puffer. On this evidence the claimant was refused a decree of specific performance, although there had been no fraud.

---

1   Such an action requires the misrepresentation to have brought about a contract between the parties.
2   (1971) 223 EG 33, 219.
3   (1788) 2 Bro CC 326.

A similar decision was reached in *Mason* v *Armitage*,[1] where the claimant, a friend of the defendant vendor, had told him that he would not buy the defendant's property at the reserve price. When the claimant subsequently bid at the sale, the defendant's puffer assumed that he did so as agent for the vendor. The puffer therefore did not puff, and the property was knocked down to the claimant at less than its reserve. Specific performance was again refused.

A more direct means of damping the sale was employed in *Fuller* v *Abrahams*[2] which concerned the sale of a barge taken in execution of a debt due from its owner. The claimant, who had built the barge, told the company that the owner had never paid him for it, whereupon no one bid against him. The auctioneer having refused to knock the barge down on a single bid, a friend of the claimant made an intervening bid, and the barge was then knocked down to the claimant at about one third of its true value. The Court of Common Pleas held that, in these circumstances, the sale could not be treated as valid.

Bidding rings

If a vendor is not permitted to inflate the price obtained at auction by secretly bidding in person, one might have thought that the law would similarly prohibit collusion among bidders to depress the hammer price by agreeing not to compete with each other. Surprisingly, perhaps, there appears to be nothing in the general law to prohibit an agreement between intending bidders at an auction, whereby only one of them bids, so as to reduce the price at which the lot is purchased. Despite an early *dictum* of Gurney B,[3] that such an agreement constituted a criminal conspiracy, the courts of equity have consistently refused to allow a vendor to avoid a contract of sale by proving that the highest bidder was party to such a 'knock-out' arrangement. Thus in *Re Carew's Estate*,[4] where land was being sold by order of the court, two owners of adjoining land agreed that only one of them should bid and that, if his bid proved successful, the land

---

1   (1806) 13 Ves 25.
2   (1821) 6 Moo CP 316.
3   *Levi* v *Levi* (1839) 6 C & P 239, 240. This view was disapproved by Park B in *Doolubdass* v *Ramloll* (1850) 7 Moo PC 239, 263, and by Bankes and Atkin LJJ in *Rawlings* v *General Trading Co* [1921] 1 KB 635, 641, 649.
4   (1858) 26 Beav 187.

should then be divided between them. The property was knocked down to this bidder for little more than the reserve price; nevertheless the sale was held to be valid.

In *Heffer* v *Martyn*[1] the claimant offered a prospective rival bidder for a plot of land £500 to refrain from bidding, provided that the claimant then obtained the land for not more than £6,300. In fact, the land was knocked down to the claimant for a mere £3,000, which was its reserve price. Despite this clear undervalue, the sale was upheld, and Lord Romilly MR said:[2]

> The intending buyers may arrange between themselves which lots they will bid for and which not, and agree not to compete with each other; and if they may do so in that case I think also they may take money for abstaining to compete as well as arrange to take one lot against another ... This is, no doubt, very hard upon a vendor, that if he combines with others to keep up the price, it should be illegal and the sale void, but that the purchaser may combine together with others to get the property at less than its value; but the real remedy in such a case as this is in fixing the reserved bidding: it is, no doubt, a very important matter, and requires very careful previous consideration.

Not only is the vendor bound in such a case to go through with the sale; the bidding agreement is itself enforceable as between the participants. In *Galton* v *Emuss*,[3] for example, where the claimant abstained from bidding for certain land in return for an option granted by the successful bidder, it was held that the option was enforceable.

The most common type of bidding agreement is one under which the spoils are divided, either in agreed proportions or by a subsequent auction among the parties (the true 'knock-out'). It is well established that, in such a case, specific performance will be available to ensure that the division is duly carried out.[4] In *Rawlings* v *General Trading Co*[5] two dealers, who attended a sale of surplus property belonging to the Ministry of Munitions, agreed that only one should bid and that the goods should be equally divided between them. The Court of Appeal, after a review of the authorities, reversed the decision of Shearman J

---

1   (1867) 36 LJ Ch 372.
2   At p 373.
3   (1844) 1 Coll 243.
4   *Chattock* v *Muller* (1878) 8 ChD 177; *Pallant* v *Morgan* [1953] Ch 43; *Du Boulay* v *Raggett* [1989] 1 EGLR 229.
5   [1921] 1 KB 365; followed in *Cohen* v *Roche* [1927] 1 KB 169.

and held the agreement to be enforceable. In a notable dissenting judgment, however, Scrutton LJ was of the opinion that the agreement, while reasonable as between the parties, was contrary to the public interest (since this was a sale of public goods) and could therefore be treated as an unlawful restraint of trade. This argument might well have pointed the way for future development of the law, but it has not been adopted in later cases. Thus in *Harrop* v *Thompson*,[1] for example, Templeman J upheld the legality of an agreement whereby a purchaser of land had induced another person not to bid against him, and thus granted a decree of specific performance against the vendor. Similarly, in *Du Boulay* v *Raggett*, where part of a country estate was being sold at auction, a number of neighbouring landowners agreed that one of their number should bid as agent for all, and that the land should then be divided in an agreed manner. This agreement was held specifically enforceable against the appointed agent.

While sales of land at auction are still governed by the legal principles described above, it is worth noting that statute has intervened in relation to the sale of goods. A 'dealer'[2] who gives or offers anything to another person in return for abstaining from bidding, or any person who accepts such an inducement from a dealer, commits a criminal offence,[3] and the seller in such circumstances is entitled to avoid the sale.[4]

---

1   [1975] 2 All ER 94. Templeman J regarded himself as bound by the *Rawlings* case.
2   Defined as someone who, in the normal course of business, attends auction sales for the purpose of purchasing goods with a view to reselling them.
3   Auctions (Bidding Agreements) Act 1927, s 1.
4   Auctions (Bidding Agreements) Act 1969, s 3.

Misinformation

An estate agent or auctioneer is normally employed by a client who wishes to sell property, and will owe the client a duty to use all reasonable care and skill in seeking to obtain the best possible price. To this end, the agent will be expected to describe the property for sale as favourably as possible in advertisements, in the particulars of sale and in direct oral communications with prospective purchasers (including, in the case of an auctioneer, bidders at the sale). This chapter is concerned with the potential legal liabilities which may arise where the agent's description of the property goes beyond what is merely 'favourable' and becomes positively untrue or, at least, misleading. Such a statement may entitle a purchaser to bring a civil action against either the agent or the vendor; moreover, the agent may in certain circumstances incur liability under the criminal law.

The basis of liability

As will be seen, the provision of information which is false or misleading may lead to civil liability under a range of headings. For the vendor, these include breach of contract and misrepresentation; for the agent, there are the torts of negligence and fraud and (possibly) an action for breach of warranty of authority. What a claimant has to prove, in order to succeed, will differ according to which type of claim is brought. However, one element is common to all: the agent must be shown to have made a false or misleading 'statement of fact'. Precisely what that means we now consider.

Statement

In most cases, a 'statement' made by one person to another will take the form of words, whether they be printed, typed, written or spoken. However, the law also treats as statements other, non-verbal, means of communication. As section 1(1)(c) of the Property Misdescriptions Act 1991 puts it:

> [A] statement may be made by pictures or any other method of signifying meaning as well as by words

While this statutory provision is of direct relevance only where a prosecution is brought under the 1991 Act, there is no doubt that the principles expressed apply equally to other types of legal action. In *St Marylebone Property Co Ltd* v *Payne*,[1] for example, auction particulars of a shop in a terrace of buildings contained a photograph, on which arrows had been drawn by the auctioneers to indicate the boundaries of the property for sale. One of these arrows was in the wrong place; it suggested that the shop included an imposing stone-surrounded doorway, whereas in fact this belonged to the adjoining property. The photograph was held to constitute both a misrepresentation and a contractual misdescription of the property for sale, notwithstanding a clause in the sale particulars stating that arrows on the photographs were for location purposes only. The purchasers were accordingly entitled to rescind the contract of sale and to recover the deposit which they had paid.

The 'statement' in the *St Marylebone* case consisted of positively tampering with a photograph. However, an unaltered photograph may also lead to liability where, though not false in an objective sense, it is nevertheless misleading.[2] In the unusual case of *Mustafa* v *Baptist Union Corporation Ltd*,[3] the claimant purchased at auction a rambling four-storey terraced house in East London, which at the time was converted into two maisonettes. The claimant inspected the property before the sale but failed to realise that, within what appeared to be the physical boundaries of the house, there was a room to which he would have no

---

1   [1994] 2 EGLR 25.
2   According to the Property Misdescriptions Act 1991, s 1(1)(b): 'a statement is misleading if (though not false) what a reasonable person may be expected to infer from it, or from any omission from it, is false.'
3   [1983] 1 EGLR 177.

access and which in fact formed part of the adjoining property. The particulars of sale accurately described those rooms which were included in the sale but made no mention of the blocked-off part of the premises. It was held that a reasonable purchaser, on seeing the photograph of the front of the house which appeared on the particulars, would assume that everything lying squarely behind it was included in the sale. It was accordingly the vendors' responsibility positively to correct this 'misdescription'.

One potential trap for agents who incorporate photographs in sale particulars is that these inevitably relate to the single moment in time at which they are taken, and may thus become misleading through subsequent changes in circumstances. The law regards a representation, once made, as continuing up to the time that a contract is entered into, and so an out-dated photograph may be a source of liability. This is what happened in *Atlantic Estates Ltd v Ezekiel*,[1] which concerned an auction sale of investment properties. A photograph attached to the particulars of one of these properties gave the impression that it was a thriving wine bar; this was perfectly true when the photograph was taken, but it rapidly became misleading when the tenant lost the licence to sell alcohol. It was held that the photograph constituted a misrepresentation, and that the purchaser was accordingly entitled to rescind the contract and to recover damages from the vendors in respect of the costs wasted in purchasing the property.

A photograph may be the most common type of non-verbal statement made to a purchaser, but it is not the only possibility. A property in the course of development is not infrequently marketed on the basis of an 'artist's impression', and this is surely a means of conveying information about the completed property. And in *Lewin v Barratt Homes Ltd*,[2] where a residential developer was prosecuted under the Property Misdescriptions Act 1991, it was held that a 'show house' of a particular type represented a statement that the developer's intention at that time was to build houses to the same design and specification.

The final type of non-verbal 'statement' is one made by conduct. In *Gordon v Selico*,[3] for example, it was held by the Court of Appeal that a landlord who deliberately covers up evidence of dry rot in a flat is

---

1   [1991] 2 EGLR 202.
2   [2000] 1 EGLR 77.
3   [1986] 1 EGLR 71

effectively representing to potential tenants that the flat is free from any such defect; he or she is thus guilty of fraud.[1]

Fact

The law governing misrepresentation requires there to be a statement of fact, since this is the only kind of statement that can be objectively false. This rules out, among other things, a promise as to future conduct, which can only give rise to liability if it forms part of a contract. In the case of *Lewin* v *Barratt Homes Ltd*, mentioned above, the developers were liable, not for breach of a promise to replicate the show house, but on the basis that they were claiming falsely that their present intention was to do so.

The 'statement of fact' requirement also means that a person cannot be held liable for what is merely a statement of opinion.[2] However, it is important to note that what appears at first sight to be an expression of opinion may, on closer examination, turn out to contain a statement of fact. This may be, first, where a person expresses an opinion which he or she does not in truth hold. This is fraud, since the person is lying about an objective fact, namely, the state of his or her opinion. Secondly, and less obviously, the law treats an expression of opinion by a person who is in possession of the relevant facts as impliedly containing a statement that those facts provide a reasonable basis for the opinion. Hence, if those facts would not, in the eyes of a reasonable person, support the opinion expressed, there is a misrepresentation. This principle is illustrated by the case of *Smith* v *Land and House Property Corporation*,[3] where an hotel was described in auction particulars as being let to 'a most desirable tenant'. The landlords had in truth experienced repeated difficulties in extracting the rent from the tenant, who was still in arrears when the particulars were drafted. The Court of Appeal held that no reasonable person, knowing the background, could have described this tenant as 'desirable'. The purchasers were therefore entitled to rescind the contract on the ground of misrepresentation.

---

1   In the absence of a deliberate cover-up, the landlord would not be liable, for there is no positive duty on a vendor or landlord to disclose physical defects in the property, even where he or she is well aware of them.
2   *Bissett* v *Wilkinson* [1927] AC 177.
3   (1884) 28 ChD 7.

One particular aspect of the opinion/fact distinction lies in the indulgence shown by the law to estate agents and auctioneers who display a degree of over-enthusiasm for the properties they are seeking to sell. In 'puffing' the property by stressing (and perhaps rather exaggerating) its good points, no liability is incurred while what is expressed is a matter of opinion rather than fact. The rather patronising attitude of the courts to the natural enthusiasm of a selling agent is well illustrated by the modern case of *Watson v Burton*,[1] where property was described as 'valuable and extensive premises . . . situated in a first-class position' and 'very suitable for development'. This description, said Wynn-Parry J,[2] was 'what I may describe, with no disrespect, as typical auctioneers' "puff"'.

There are many examples in the law reports of statements falling on either side of the line which separates 'puff' from misdescription (and, therefore, legal liability). In *Scott v Hanson*,[3] land described as 'fourteen acres of uncommonly rich water-meadow land' was in fact imperfectly watered. It was held that, provided the description of it as a water meadow was accurate, its 'richness' was merely a matter of opinion. Similarly, in *Dimmock v Hallett*,[4] a statement that land was 'fertile and improvable' was disregarded, even though part of the land had been abandoned as useless. On the other hand, where a farm was described as being 'in a high state of cultivation', the purchaser was awarded compensation for its very impoverished state.[5] And in *South Western General Property Co Ltd v Marton*,[6] it was held that the description 'development land' was a misrepresentation when applied to a plot of land, for the development of which planning permission had been repeatedly refused.

It is perhaps in relation to houses that estate agents and auctioneers wax mostly lyrical; Sachs J has described purchasers' 'recurrent joy of comparing the realities of some property with the benign description in the estate agent's particulars'.[7] However, a purchaser who challenges their description must show more than a difference of opinion in order to obtain a remedy. Thus it is merely a 'puff' to describe property as 'a

---

1   [1956] 3 All ER 929.
2   At p 931.
3   (1826) 1 Sim 13.
4   (1866) 2 Ch App 21.
5   *Dyer* v *Hargrave* (1805) 10 Ves 505.
6   [1982] 2 EGLR 19.
7   *Goding* v *Frazer* [1966] 3 All ER 234, 238.

residence fit for a family of distinction'[1] or a 'substantial and convenient dwelling-house'.[2] Where, however, a house is described as 'in good repair'[3] or 'not damp'[4] these are treated as statements of fact and, if they are untrue, the purchaser is entitled to recover compensation from the vendor or to avoid the sale. In *Registered Holdings Ltd* v *Kadri*,[5] the defendant successfully bid for what was described from the rostrum as a 'nice house' in Battersea, only to discover that his standards of 'niceness' went some way beyond those of the auctioneer. In holding that the vendors were not entitled to enforce the sale, Goff J said:[6]

> By no stretch of the imagination can a house the whole of which is subject to a compulsory purchase order, albeit provisional, and the basement of which, including the kitchen, is subject to a closing order be described as 'nice'. In my view, this goes beyond the realm of puff and is a misrepresentation.

Liability of the vendor
Breach of contract

A vendor will be guilty of a breach of contract, sometimes referred to as 'misdescription', where property does not match up to its description in the contract of sale. In such circumstances the purchaser will be entitled to an award of damages assessed on a contractual basis,[7] and, if the misdescription is a substantial one, to set aside the contract.

The extent to which the activities of estate agents or auctioneers are likely to impose liability of this kind upon a vendor is limited. While auction particulars frequently form part of the contract of sale, this is not true of estate agents' sale particulars (which, indeed, routinely contain a provision expressly denying their contractual status), nor of advertisements issued by agents. Nevertheless, the courts have in a

---

1   *Magennis* v *Fallon* (1829) 2 Moll 561.
2   *Johnson* v *Smart* (1860) 2 Giff 151.
3   *Dyer* v *Hargrave* (1805) 10 Ves 505.
4   *Strangways* v *Bishop* (1857) 29 LT (OS) 120.
5   (1971) 222 EG 621.
6   At p 625.
7   That is to say, the sum awarded will seek to put the purchaser into the financial position he or she would have been in if the property had really been as it was described.

number of cases been faced with a claim to the effect that something said or written by an auctioneer or estate agent should be treated as a 'warranty', that is a contractual promise made on behalf of the vendor.

In *Payne* v *Lord Leconfield*,[1] a mare which was being sold by auction was observed to have a discharge from her nostrils. The auctioneer, in the claimant's hearing, said: 'You need not be afraid. The mare comes from Lord Leconfield; she has only got a cold upon her, and I shall sell her as only having a cold.' The mare was knocked down to the claimant but was found to be suffering from chronic glanders, as a result of which the local authority ordered her to be shot. The claimant thereupon sued the vendor for damages for breach of this warranty, but his claim was rejected by the Court of Queen's Bench. It was held that an auctioneer, who is simply an agent to sell, has no implied authority to warrant goods and must therefore approach the principal to seek express authority.

Having refused to imply authority in the case of an auctioneer, whose function includes the making of a contract of sale, it would be surprising to find the courts adopting a more expansive view of estate agents. Nor, indeed, have they done so: in *Lawrence* v *Hull*,[2] where an estate agent's newspaper advertisement described a house as being 'in perfect order', Rowlatt J held that this could not be treated as a warranty, so as to entitle the disappointed purchaser to damages. His lordship said 'it might be that in equity a vendor whose agent had made a misrepresentation of this kind would not be able to enforce specific performance, but that was a very different thing from holding that the representation was a warranty'.

In *Hill* v *Harris*[3] the defendant, who was the tenant of premises on which he carried on the trade of boot and shoe maker and dealer, instructed estate agents to find a sub-tenant for the premises. The claimant wanted the premises for a confectionery and tobacco business and, although the head lease prohibited such a use, the estate agents told him that it would be all right. When the head lessor prevented the claimant from using the premises in this way, the claimant sought damages for breach of an implied collateral warranty that, if he took a lease of the premises, the defendant would guarantee that he could use them as he intended. The Court of Appeal held that no collateral warranty existed and Diplock LJ said:

---

1   (1882) 51 LJQB 642.
2   (1924) 41 TLR 75.
3   [1965] 2 QB 601.

It is, I apprehend, clear law that the ostensible authority of an estate agent invited to find a purchaser for premises or a lessee for premises, does not extend to entering into any contractual relationship in respect of the premises on behalf of the person instructing him. It may well be that he has authority to make representations as to the state of the premises, but representations are a very different matter from warranty.

Misrepresentation

A misrepresentation is a false or misleading statement of fact which induces someone to enter into a contract. If proved, it entitles the party misled to rescind the contract and/or claim damages.[1] In most cases, the statement is made directly by one contracting party to the other, but the position is the same where the offending statement is made by a contracting party's authorised agent.[2]

That a vendor may be held responsible for an auctioneer's misrepresentation (that is to say, that it lies within the auctioneer's implied authority) is established by the case of *Smith* v *Land and House Property Corporation*.[3] Auction particulars there described an hotel as let to 'a most desirable tenant', when in fact the tenant was in arrear with the rent. The purchasers were held entitled to rescind the contract on the ground of misrepresentation, and Baggallay LJ, said:[4]

It is said that these are words of course put in by the auctioneer, but I hold it to be the duty of a vendor to see that the property is not untruly described, and I cannot hold him to be excused because a description which the property will not bear has been inserted by the auctioneer without his instructions.

Two years earlier, in the case of *Mullens* v *Miller*,[5] the courts had for the first time considered the position of a vendor whose estate agent makes a misstatement. The claimant there employed an estate agent to find,

---

1   Which remedies are available in any given case depends upon whether the misrepresentation is made fraudulently, negligently or innocently.
2   If the agent is not authorised to make the relevant statement, the vendor cannot be held responsible, which means that the innocent party cannot rescind the contract. However, the agent may be personally liable for breach of warranty of authority: see p 219.
3   (1884) 28 ChD 7.
4   At p 13.
5   (1882) 22 ChD 194.

first, a tenant and, later, a purchaser for his warehouse. The agent induced the defendant to offer £750 by making wildly exaggerated claims as to the value of the warehouse and by naming, falsely, various persons who were said to be eager to lease it. When the truth was discovered, the defendant refused to complete the purchase; the claimant then brought an action for specific performance of the contract, claiming that, since the agent's statements were made without any form of authority whatsoever, they could not prejudice the rights of the seller.

In refusing to award specific performance, Bacon V-C dealt with this contention as follows:

> A man employs an agent to let a house for him; that authority, in my opinion, contains also an authority to describe the property truly, to represent its actual situation, and, if he thinks fit, to represent its value.

A similar result was reached in *Wauton* v *Coppard*,[1] although the question of authority does not appear to have been specifically raised in that case. There, an auctioneer, selling the defendant's leasehold property by private treaty (and therefore in the same position, legally speaking, as an estate agent) misrepresented the effect of certain restrictions upon the permitted use of the property under the lease. The court, referring throughout to the statements complained of as being made by the defendant's agent, awarded the claimant rescission of the contract and the return of the deposit with interest.

These examples illustrate clearly the liability to which a client may be subjected by an agent's misrepresentation, and it might well be supposed in such circumstances that the client could recover this loss by taking action against the agent. This assumption was proved accurate in *Whiteman* v *Weston*,[2] where the defendant, an estate agent instructed to find a tenant for the claimant's house, told a music teacher that he could use the property as a school of music. This use in fact constituted a breach of covenant and the claimant was in due course compelled, by complaints from the neighbours, to take legal action against his tenant. This action for breach of covenant failed, because the tenant was entitled entirely to rely on the assurances given by the estate agent, and the landlord then succeeded in recovering the costs of his abortive action from the agent by way of damages for breach of duty.

---

1   [1889] 1 Ch 92.
2   [1900] *The Times*, March 15. See also *Power* v *Atkins* [1921] NZLR 763, and its sequel, *Atkins* v *A D Kennedy & Co Ltd* [1921] NZLR 977.

The effect of an estate agent's misrepresentation fell to be considered in two later cases concerning commission.[1] In each case, commission was payable on the signing of a binding contract, and in each case the purchaser, incensed at a misstatement made by the estate agent, refused to complete the contract he had signed. Both estate agents failed in their claims for remuneration since, as was admitted on all sides, the misrepresentations made these contract unenforceable against the purchasers.

Had the line of judicial comment stopped at this point, it would have appeared settled beyond any reasonable argument that clients are responsible for statements made by their estate agents. However, some doubt was cast by certain remarks of Roskill LJ in *Gosling* v *Anderson*.[2] In that case the claimant bought a flat from the defendant on the faith of an assurance from an estate agent that planning permission had been obtained for the erection of a garage. Graham J at first instance,[3] impressed by the fact that the defendant had left all negotiations in the hands of the estate agent, said that he 'had ostensible authority to finalise matters and in particular to answer questions and give assurances about the property and the position in regard to planning permission'. Nevertheless, the claimant failed in her claim for damages under section 2(1) of the Misrepresentation Act, 1967 since, in the eyes of the court, she had not relied on the statement to a sufficient extent in deciding to purchase.

This latter finding was reversed by the Court of Appeal, so that the claimant was, after all, awarded damages. Only Roskill LJ made any mention of the question of an estate agent's authority, saying:

> The learned judge spoke of the agent having ostensible authority from the defendant. In my view he was wrong in so stating. The relevant modern authorities on ostensible authority were not cited to him. The agent had actual authority to make the representation relied on, and the defendant is bound by that representation.

Since Roskill LJ did not identify the 'relevant modern authorities', it is difficult to establish exactly why he denied the existence of any ostensible authority in this case. It may indeed be that his reference to

---

1   *Gregory* v *Fearn* [1953] 2 All ER 559; *Peter Long & Partners Ltd* v *Burns* [1956] 1 WLR 413, 1083.
2   (1972) 223 EG 1743.
3   (1971) 220 EG 1117.

'actual authority' was intended to include implied authority, in which case his judgment would be in line with the earlier cases to which we have referred. If, however, he meant to suggest that a client will be liable only where the agent has been expressly authorised to make the statement in question, it is submitted that this is incorrect. For safety's sake it should be assumed that statements made by an estate agent may be a source of liability for the client as well as for the agent personally.

In the vast majority of cases involving misrepresentation by an agent, the misrepresentation had been made innocently or, at worst, negligently. Occasionally, however, the agent is guilty of fraud, and the position of the vendor in such a case merits special attention. As seen in chapter 1, the vendor may in principle be held liable for a statement made within the agent's authority, even though the agent has deliberately lied. There is, however, one particular problem. This is where the agent makes a false statement innocently, while the principal, who knows the truth, is unaware that the statement is being made.[1] In *Cornfoot* v *Fowke*,[2] an agent told the prospective tenant of a house that there was nothing objectionable about it. In truth, as the principal (though not the agent) knew all too well, the house was next door to a brothel. The tenant sued the principal for fraud, but the action failed. A similar point arose in the case of *Armstrong* v *Strain*,[3] which concerned the sale of a much-underpinned bungalow. The vendor did not mention this fact to the estate agents, who innocently misrepresented its condition to the purchaser. Devlin J, whose decision was affirmed by the Court of Appeal, concluded that neither vendor nor estate agent was liable for fraud, since 'you cannot add an innocent state of mind to an innocent state of mind and get as a result a dishonest state of mind'.[4]

Avoidance of liability

Where there is a risk of liability for misrepresentation, there is frequently an attempt to exclude that liability by means of a contract term or an extra-contractual notice. An account of the legal rules

---

1   If the principal intends that the agent should make the offending statement, he is personally guilty of fraud: *Ludgater* v *Love* (1881) 44 LT 694.
2   (1840) 6 M & W 358.
3   [1952] 1 KB 232.
4   1951] 1 TLR 856, 872.

governing such devices is beyond the scope of this book; they apply to exemption clauses generally, not just to those where an offending statement is made by the agent of a contracting party. Nevertheless, it is relevant to consider one particular method by which vendors have in recent years sought to avoid liability for false statements made by an estate agent or auctioneer on their behalf. This consists of an announcement to the world at large that the agent concerned has no authority to that effect, so as to preclude any implication of authority which might otherwise be made.

This device first came before a court in the case of *Overbrooke Estates Ltd* v *Glencombe Properties Ltd*,[1] where the defendants, who had purchased property at an auction, refused to complete the purchase on the ground that the auctioneers, three days before the sale, had made a material misrepresentation about the property. The claimant vendors, while not denying this allegation, sought to rely upon a general condition of sale in the following terms:

> The Vendors do not make or give and neither the Auctioneers nor any person in the employment of the Auctioneers has any authority to make or give any representation or warranty in relation to these properties.

It was held by Brightman J that the defendants, who must be taken to have bid with knowledge of the condition, were fixed with notice of the auctioneers' lack of authority and were therefore bound by the sale. Further, the condition was held not to be 'a provision which would exclude or restrict ... any liability' within the meaning of section 3 of the Misrepresentation Act 1967, so that it did not have to satisfy the test of reasonableness imposed by that statute.

Overbrooke v *Glencombe* was approved and, indeed, extended, by the Court of Appeal in *Collins* v *Howell-Jones*.[2] Estate agents there, acting on the specific instructions of a client, showed a prospective purchaser certain drawings of a property for sale which were misleading (the drawings implied that a two-storey extension could be built at the rear of the property, whereas in truth such an extension would have infringed the rights to light of adjoining landowners). The purchaser, having acquired the property, sought compensation from the client on the ground of misrepresentation, but the Court of Appeal held that a 'no authority' clause in the estate agents' particulars of sale operated to

---

1   [1974] 3 All ER 511.
2   [1981] 2 EGLR 108.

defeat his claim. In so deciding, the court held that the *Overbrooke* principles applied, even though the clause in this case did not form part of the contract (the sale particulars themselves made this clear) and even though the agents here had been expressly authorised to make the offending statement.

The Court of Appeal in *Collins* v *Howell-Jones* did not have to consider the effect (if any) which section 3 of the Misrepresentation Act 1967 might have upon the clause in question, for section 3, as amended by section 8 of the Unfair Contract Terms Act 1977, now applies only to exemption clauses which form part of a contract. However, the view of Brightman J mentioned above, that an authority-denying clause is in any event outside the scope of the statute, must now be regarded a little more doubtfully in the light of *South Western General Property Co Ltd* v *Marton*.[1] That case concerned the sale of a plot of land at auction, so that the contract which was made on the fall of the hammer included the conditions of sale. The purchaser claimed to rescind the contract on the ground of a misrepresentation contained in the auctioneers' catalogue, whereupon he was faced with a variety of exemption clauses set out in the conditions of sale, including one of the kind now under discussion. It was held by Croom-Johnson J that all these clauses should be struck out as failing to satisfy the statutory requirement of 'reasonableness'; the possibility that the authority-denying clause was not subject to this test does not appear to have been considered or even mentioned.

Liability of the agent
Negligence

The question whether an estate agent or auctioneer, in disseminating information on behalf of a vendor, owes a personal duty of care to a potential purchaser, is one on which there is relatively little English authority.[2] Moreover, such case law as there is has not settled the agent's position quite as clearly as might have been hoped.

The first case in which this question fell to be considered was *Computastaff Ltd* v *Ingledew Brown Bennison & Garrett*,[3] where the claimants had retained a firm of estate agents (the second defendants)

---

1   [1982] 2 EGLR 19.
2   However, Commonwealth jurisdictions generally favour the imposition of such a duty.
3   [1983] 2 EGLR 150.

to seek office accommodation for them to rent. The agents found suitable premises and, taking the information from particulars drawn up by the landlords' estate agents, told their clients that the rateable value of the property was £3,305. The correct figure was in fact £8,305 but, when the discrepancy was noticed (the landlords' solicitors quoted the correct figure in reply to a preliminary enquiry), the claimants' solicitors and the two firms of agents investigated the matter in such a negligent fashion that the error was simply repeated. The claimants duly took the lease and, on finding that the burden of the rates was far heavier than they had been led to expect, sued their own agents and solicitors in negligence. These defendants joined the landlords' agents as third parties under the Civil Liability (Contribution) Act 1978, seeking a contribution from them towards any damages payable. Such a contribution could only be ordered where the third parties would, if sued, have been directly liable to the claimants in the tort of negligence. Agreeing that they would have been so liable, McNeill J held that, as between the claimants' advisers, the solicitors were 40% and the estate agents were 60% to blame; these defendants were each entitled to recoup one-half of what they had to pay from the landlords' agents.

The most important decision in this area is undoubtedly that in *McCullagh* v *Lane Fox & Partners Ltd*,[1] which concerned the sale of a substantial riverside residential property in West London. In showing the claimant round the property on a Saturday morning, a senior member of the defendant estate agents orally confirmed what was stated in the sale particulars, namely, that the overall plot size was 0.92 acres. On the Sunday, the claimant made an offer for the property and, on being told that there was a rival purchaser, increased his offer and promised to exchange contracts on the following day. After contracts were exchanged, the claimant discovered that the plot was in fact only 0.48 acres in size. Unable, because of terms in the contract of sale, to take action against the vendor in respect of this misrepresentation, the claimant sued the estate agents, alleging that his reliance on their negligent statement had led him to pay substantially more than the property was worth.

The trial judge, having referred to the *Computastaff* case and to the general case law on negligent statements, appeared to think that the agents owed no duty of care to the claimant when they first gave him the inaccurate information, because they would not then have

---

1   [1994] 1 EGLR 48; reversed [1996] 1 EGLR 35.

realised that the claimant would enter into a contract of purchase, without first having the property surveyed or otherwise investigated by his own professional advisers. However, he held that the agents' statement should be treated as continuing until the moment that contracts were exchanged and that, by the time this was done, they were aware of the claimant's reliance; they accordingly then owed a duty of care to the claimant to correct their error. Even so, the claim failed because, in the judge's view, the property had in fact been worth what the claimant paid for it.

On appeal, the finding on value was reversed; the claimant was held to have suffered a loss of £75,000. However, this ruling was of merely academic interest, since the Court of Appeal also reversed the judge's decision on the basic question of the estate agents' duty of care. According to Sir Christopher Slade and Nourse LJ, this was because the idea of a duty to correct an error, while well established in the context of misrepresentation, was not applicable to actions in the tort of negligence. Hobhouse LJ was prepared to accept that there could be a duty to correct, but held that, by the time this was breached, the claimant was in possession of the defendants' sale particulars, which contained a clause to the effect that all statements about the property were made 'without responsibility'. And, since this disclaimer satisfied the statutory test of reasonableness,[1] it was effective to protect the agents from liability.

It seems clear from the judgments of the Court of Appeal in this case that where, as is normal, an estate agent can expect a purchaser's solicitors and surveyors to check out the property before a contract is entered into, the agent will owe no duty of care in respect of what is said or written about the property. In effect, the agent does not expect his or her words to be relied upon by the purchaser. However, and notwithstanding the strength of the opinions expressed by the Court of Appeal, it is clearly possible for the evidence in a particular case to show that an estate agent has assumed responsibility for what he or she says, in a way sufficient to justify a court in recognising a duty of care. An example is the case of *Duncan Investments Ltd* v *Underwoods*,[2] which arose out of the sale by receivers of a portfolio of 16 residential

---

1   Unfair Contract Terms Act 1977, s 2.
2   [1997] PNLR 521. See also the Irish case of *McAnarney* v *Hanrahan* [1993] 3 IR 492, where an auctioneer was held liable in negligence for misinforming the claimant that a leasehold property had been withdrawn from a recent sale following a bid of £54,000, and that the freehold interest could be acquired for about £3,000.

properties. The claimants, who were considering the purchase of the entire portfolio, asked the defendant estate agents, who were handling the sale, what would be an appropriate asking price on a resale of each property, and what price each could reasonably expect to achieve on a sale within six months.

It was held by Lloyd J that, in giving advice on these matters, the defendants owed the claimants a duty of care. They knew that the claimants would rely on their advice, and that they were unlikely to obtain other independent advice. The judge held, further, that a clause in the defendants' sales particulars stating that they had no authority make or give any representation or warranty did not provide them with a defence. This clause was effective only to protect the vendors, by making it clear that their agents were not authorised to make representations on their behalf; it had no bearing on the agents' own liability for their negligent statements.

On appeal,[1] the defendants did not specifically challenge the judge's ruling on duty of care, although the Court of Appeal considered the disclaimers and agreed with the judge's conclusion that the 'no authority' clause provided no defence. However, the court made a substantial reduction in the damages payable, ruling that the assessment of the claimants' loss should have been based on the total value which the properties would have had if purchased individually, rather than the discounted price at which they could have been purchased as a portfolio.

Other forms of liability

In considering whether a misstatement by an estate agent or auctioneer may lead to legal liability in other ways, two clear points emerge. First, an agent who makes a statement which he or she knows is untrue commits the tort of deceit (or fraud) and is personally liable for this, whether or not the client can also be held responsible.[2] Second, the Misrepresentation Act 1967 applies only to statements made by or on behalf of one contracting party to another; it follows that an agent who makes an offending statement cannot incur personal liability under that Act.[3]

---

1   [1998] PNLR 754.
2   See p 77.
3   *Resolute Maritime Inc* v *Nippon Kaiji Kyokai* [1983] 2 All ER 1.

One possible cause of action against an estate agent or auctioneer, albeit one which does not appear to have been tried in an English court, is an action for breach of warranty of authority. This is a well-established form of liability for any agent, which arises where the agent exceeds the authority given by the client, with the result that the client is not bound to the third party by the agent's act. In the present context, one might argue that, where an agent's statement about property is not binding on the client, the agent can be held directly liable to the purchaser for purporting to make it on the client's behalf.

In *Payne* v *Lord Leconfield*,[1] where it was held that an auctioneer had no implied authority from the client to give a contractual guarantee about a horse, Grove J stated in passing that the auctioneer could have been made personally liable for breach of warranty of authority. However, there is an obstacle, arising out of modern sale documentation, which should not be overlooked. An action for breach of warranty of authority will not in principle succeed where the third party is aware that the agent in fact has no authority, and today's sale particulars or conditions routinely deny any authority of the estate agent or auctioneer to make statements. The use of such clauses may well mean that this cause of action will seldom be available to purchasers.

It remains to consider one specific type of misstatement, namely, an assurance that a sale by auction will be held at a particular time and place. It appears that, provided that such a statement is made in good faith, it imposes no liability on the advertiser to hold the sale. In *Harris* v *Nickerson*,[2] an auctioneer advertised a sale of office furniture to be held at Bury St Edmunds. The claimant, having obtained a commission to buy certain lots, travelled from London, but the lots were withdrawn. It was held that the claimant could not maintain an action against the auctioneer for the cost of his abortive journey.

The good faith of the auctioneer in that case was an important factor, as was made clear by the Court of Queen's Bench in *Richardson* v *Silvester*,[3] a case decided some six months later. There the defendant, without any authority, advertised a large farm to let. The claimant, who was eager to take a tenancy of the farm, incurred expense on the faith of the advertisement. It was held that an action for damages in the tort of deceit would lie on these facts.

---

1   (1882) 51 LJQB 642.
2   (1873) LR 8 QB 286.
3   (1873) LR 9 QB 34.

Criminal liability

As explained in chapter 7, the Government in the 1980s intended to extend the Trade Descriptions Act 1968 to cover false or misleading statements about land and buildings and, as an interim measure, to designate the making of such statements an 'undesirable practice' under the Estate Agents Act. This intention, however, was finally overtaken by events; a Private Member's Bill entitled the Estate Agents (Property Misdescriptions) Bill was introduced into Parliament and this, with Government support, duly passed into law as the Property Misdescriptions Act 1991.[1]

Scope of the Act

The area of application of the Property Misdescriptions Act is for the most part similar to that of the Estate Agents Act, although it extends further in two important respects. According to section 1(1) of the 1991 Act, what is covered is any 'false or misleading statement ... made in the course of an estate agency business or a property development business, otherwise than in providing conveyancing services'.

Estate agency business

The primary target of the Property Misdescriptions Act is undoubtedly the world of estate agency. In seeking to define this, section 1(5)(e) provides in effect that a statement shall be treated as made 'in the course of an estate agency business' for the purposes of the 1991 Act if, and only if, the making of that statement would also constitute 'estate agency work' within the meaning of section 1(1) of the Estate Agents Act 1979. To this extent, therefore, everything said in chapter 7 about the scope of the 1979 Act is also relevant to potential liability for property misdescriptions.[2]

---

1 The change of name came about when the Standing Committee widened the scope of the Bill to include statements made by property developers and builders (in marketing what they have built) as well as by estate agents.

2 See pp 270–285.

Solicitors

As will be seen,[1] the Estate Agents Act provides a specific exemption for things done by a practising solicitor (or by any employee of a solicitor) 'in the course of his profession'. The Property Misdescriptions Act confers no such immunity, and thus includes within its definition of 'estate agency business' anything which would have been caught by the 1991 Act were it not for the solicitor's exemption. As a result, any solicitor who provides estate agency services will be subject to the 1991 Act.

However, the extent of this provision should not be over-estimated. There is no intention to subject solicitors generally to the Property Misdescriptions Act, but only those who practise estate agency. The Act achieves this by excluding statements made in the course of providing 'conveyancing services', which are defined by section 1(5)(g) as 'the preparation of any transfer, conveyance, writ, contract or other document in connection with the disposal or acquisition of an interest in land, and services ancillary to that'.

Property development

The most important difference between the coverage of the Estate Agents Act and that of the Property Misdescriptions Act lies in the fact that the latter governs property developers and builders as well as estate agents. The Act applies to a statement made in the course of a 'property development business' (which is defined by section 1(5)(f) as 'a business ... concerned wholly or substantially with the development of land'), provided that the statement is made 'with a view to disposing of an interest in land consisting of or including a building, or a part of a building, constructed or renovated in the course of that business'.

The latter part of this definition serves to rule out those businesses which merely deal in property, rather than developing it. On the other hand, builders and developers who sell houses and flats directly (for example from a sales office on a housing estate) are certainly included.

---

1   P 277.

The basic offence

The intention behind the Property Misdescriptions Act 1991, which came into force on 4 April 1993, was to create a new form of strict liability criminal offence arising out of statements concerning property. The operative provision is section 1(1), which provides:

> Where a false or misleading statement about a prescribed matter is made in the course of an estate agency business or a property development business, otherwise than in providing conveyancing services, the person by whom the business is carried on shall be guilty of an offence under this section.

It is important to appreciate that this criminal liability is the only sanction for a breach of the 1991 Act. Section 1(4) provides specifically that:

> No contract shall be void or unenforceable, and no right of action in civil proceedings in respect of any loss shall arise, by reason only of the commission of an offence under this section.

It follows that, while it may be a complaint from an aggrieved purchaser which leads trading standards officers to prosecute an estate agent under the 1991 Act, the purchaser will derive no material benefit from the prosecution. A claim for damages or rescission of the contract of sale can only be made in accordance with the principles described earlier in this chapter.

The fact that there is no civil remedy for breach of the Property Misdescriptions Act makes it easier to understand what at first appears a somewhat surprising point. In proceedings under this Act, it is unnecessary for the prosecution to allege or prove that any particular person was misled by the misstatement in question or suffered any kind of loss by relying on it.[1] The offence is committed as soon as a false or misleading statement is made, for example when an advertisement is published or sales particulars are made available to potential purchasers.

---

1   *Dacre Son & Hartley Ltd* v *North Yorkshire Trading Standards* (2005) 169 JP 59.

Statement

As noted at the start of this chapter, the definition of 'statement' contained in section 1(5)(c) of the Act includes, not only written and spoken words, but also pictures and 'any other method of signifying meaning'. In the case of *Lewin* v *Barratt Homes Ltd*,[1] this extended definition was applied to both a show house and a photograph of a house of the same design (known as the 'Maidstone'). It was held that, notwithstanding warning notices attached to the photograph, the message conveyed by both the show house and the photograph was that it was the intention of the defendant developers to build all Maidstone-type houses to the same design. This was not in fact the case; the defendants were already aware that, because of difficulties with planning permission, the external form of the houses would have to be changed. It was held by the Divisional Court that the defendants were guilty of an offence under section 1 of the 1991 Act.

Notwithstanding the width of this provision, it should be emphasised that liability in all cases depends upon the making of something which can be described as a 'statement'. The Act does not impose any positive obligation of disclosure, even where an estate agent actually knows something detrimental about the property (for example that a previous survey has revealed serious defects).

False or misleading

Section 1(5)(a) provides that a false statement is one which is 'false to a material degree', which means that utterly trivial inaccuracies will not be a source of liability. As to what will be regarded as 'material', the Act gives no further guidance, but it can safely be assumed that this will take into account all the circumstances of the case, including such factors as the type of property involved (such as residential or commercial) and the class of purchaser (such as private or business).

Of course, words or pictures which are literally true may still convey a wrong impression, and these will fall within the Act if they are 'misleading'. According to section 1(5)(b), this will occur where 'what a reasonable person may be expected to infer from [a statement], or from any omission from it, is false'. To this extent only, then, liability can arise from what is not said as well as what is. A claim that 'the

---

1   [2000] 1 EGLR 77.

property enjoys extensive views over open country' may be perfectly true; but failure to mention the industrial estate which blocks the entire horizon on one side will render the statement as a whole a misleading one.

A somewhat controversial application of the 'half truth' principle appears from the case of *Enfield LBC* v *Castles Estate Agents Ltd*,[1] where a property for sale was advertised by the defendants as a 'four bedroom extended semi-detached house, plus one bedroom bungalow'. The sale particulars also referred to a 'one bedroom bungalow located in the grounds of the main dwelling'. When it was subsequently discovered that the 'bungalow' did not have the benefit of planning permission for residential use, the estate agents were prosecuted under the Property Misdescriptions Act. The magistrates' court held that to describe the building in question as a 'bungalow' was misleading, since a reasonable person might be expected to infer from this that it could lawfully be used as a residence. The magistrates in fact acquitted the agents under the defence of 'due diligence', and their decision to do so was upheld on appeal by the Queen's Bench Divisional Court.[2] However, the Divisional Court was not called upon to express an opinion as to whether the magistrates had been correct in giving this meaning to the word 'bungalow'. This is unfortunate, since it must surely be at least arguable that the word means no more than 'a single-storey building used for residential purposes', without any inference as to the lawfulness or otherwise of that use.

The approach adopted by the magistrates in *Enfield LBC* v *Castles* may be contrasted with that of the Divisional Court in the later case of *Lancashire CC* v *Buchanan*.[3] The defendant, a solicitor and estate agent, drew up sale particulars describing a property as having a garden at the side which was excellent, good sized and well fenced. It transpired that the vendor of the house was not the registered owner of the garden, although there was ample evidence to suggest that the land in question had been fenced in and used as a garden by successive owners of the house for at least 30 years. When approached by a trading standards officer, the defendant wrote setting out this evidence, though he then refused to comply with the officer's further requests to have a recorded interview and to see the defendant's file

---

1   [1996] 2 EGLR 21.
2   See p 229.
3   [2007] EWHC 3194 (Admin).

about the property. The defendant there was charged with intentionally obstructing a trading standards officer, a charge which required the officer to show that she had reasonable grounds for suspecting that an offence under section 1 of the Act (that is, the making of a false or misleading statement) had been committed. It was held that the officer could have no such reasonable suspicion since, in advertising a property for sale, an estate agent is making no representation as to the vendor's title, and any reasonable person would expect issues of title to be dealt with as part of the normal process of conveyancing. It followed that the statement to the effect that the property for sale included a garden did not carry any inference that the vendor actually owned that garden.

One particular way in which a statement can mislead is by being out of date. Under the law of misrepresentation, a person who makes a statement and then discovers it to be untrue is under a legal obligation to inform the person to whom it was made before any contract is entered into.[1] The Property Misdescriptions Act does not impose a continuing responsibility of this kind, but there is no doubt that an estate agent who continues to issue sales particulars after a change of circumstances has rendered them inaccurate may be guilty of an offence.[2]

Prescribed matter

The 1991 Act applies only where a statement is made about a 'prescribed matter', and this is defined by section 1(5)(d) to mean 'any matter relating to land which is specified in an order made by the Secretary of State'. The only such Order so far issued, the Property Misdescriptions (Specified Matters) Order 1992, contains a list of 33 items which provides virtually comprehensive coverage of both the physical and the legal aspects of property being sold. As to the former, the Order mentions such matters as measurements or sizes, form of construction, condition, environment and survey results. Examples of the latter include lease terms, rent, service charges, rates, easements and

---

1   See *With* v *O'Flanagan* [1936] Ch 575.
2   Indeed, the first recorded prosecution and conviction under the Act concerned an agent who failed to alter the price of a house on a card in his window, after the vendor had instructed him that it was to be increased: *Norfolk CC Trading Standards Department* v *Sinclair* (1993).

planning matters. Other matters which are specified, but which do not fall clearly into either of the above categories, include the property's age and history, its price and the length of time it has been on the market.

Disclaimers

Although the Property Misdescriptions Act makes no explicit reference to disclaimer clauses, they are worth a separate mention, if only because such clauses, in one form or another, are routinely found in estate agents' particulars of sale. The Act's reticence on this matter means that it will be left to the courts to determine the extent, if any, to which these clauses may serve to exclude or restrict liability. As to what they are likely to decide, it may be confidently asserted that an exemption clause of the 'no liability is accepted' kind will be of no effect whatsoever; it is simply not possible to deny one's liability for criminal offences in this way.

As to those 'disclaimers' which do not seek in so many words to deny responsibility, but attempt rather to draw the sting from what would otherwise be a false or misleading statement, the legal position is less clear. Estate agents' particulars have long carried paragraphs stating, for instance, that all statements contained therein are matters of opinion only, or that all measurements are mere approximations, or that accuracy cannot be guaranteed. Can it be said that, by denying any intention on the agent's part to assert positive facts, they effectively prevent anything in those particulars from constituting a 'false or misleading statement'?

The answer to this question, it is tentatively suggested, is that matters which are essentially factual in their nature cannot be turned into matters of mere opinion by a statement that this is all they are. Nor will a refusal to 'guarantee accuracy' mean that a court is precluded from holding a person liable for making an inaccurate statement.[1] Moreover, even if such terms might prove useful in relation to statements which truly are matters of opinion (such as a property's decorative state) or approximations (such as the area of an irregularly shaped garden), they can never amount to a licence to commit fraud. Thus, for example, an estate agent who has accurate measurements of a property cannot systematically exaggerate them under cover of a clause which states that they are mere approximations.

---

1   See *May* v *Vincent* [1991] 1 EGLR 27.

Even within their potential field of operation, it appears likely that disclaimers, to be effective, will be required to satisfy the tests laid down by the courts in cases arising under the Trade Descriptions Act 1968. In short, the question to be asked is whether the effect of a false or misleading statement on the mind of the person to whom it is made has been neutralised by an express disclaimer or contradiction of its message. As explained in the leading case of *Norman* v *Bennett*:[1]

> To be effective any such disclaimer must be as bold, precise and compelling as the trade description itself and must be as effectively brought to the notice of any person to whom the goods may be supplied. In other words, the disclaimer must equal the trade description in the extent to which it is likely to get home to anyone interested in receiving the goods.

Incidence of liability

Interestingly, it is not the actual making of a false or misleading statement which is defined as an offence under the 1991 Act. Instead, section 1(1) provides that, where an offending statement is made in the course of an estate agency or property development business, it is the person by whom that business is carried on who is guilty of an offence.[2] This policy of targeting principals is seen further in section 4(1), which provides that, where the business in question is carried on by a company, liability also attaches to any actual or *de facto* director, manager, secretary or other similar officer who has contributed by 'connivance or consent' to the commission of the offence.

The only other person who may be prosecuted under the Act is an employee through whose act or default an offending statement is made.[3] Where an employee is at fault in this way, section 1(2) makes clear that proceedings may be brought against the employee, whether or not the employer is also prosecuted.

---

1   [1974] 3 All ER 351.
2   *Dacre Son & Hartley Ltd* v *North Yorkshire Trading Standards* (2005) 169 JP 59.
3   Again the wording can impose liability, not only for making a statement, but also for causing one to be made (for example by failing to check sales particulars before publication).

The defence of due diligence

The criminal offence created by section 1 of the Property Misdescriptions Act 1991 is in principle one of strict liability, in that the prosecution is not required to prove either intent to deceive or even any negligence on the part of the defendant. However, section 2(1) provides that it shall be a defence for a person prosecuted 'to show that he took all reasonable steps and exercised all due diligence to avoid committing the offence'.[1]

There are strong echoes here of the defence under section 24 of the Trade Descriptions Act 1968 and, while the analogy cannot be pushed too far (for example because the earlier provision contains some additional elements), there is reason to suppose that the courts will pay attention to cases decided under the 1968 Act when called upon to interpret the later provision. Assuming that this is indeed the case, it is likely that 'reasonable steps' and 'due diligence' will be treated as cumulative requirements and, moreover, as demanding evidence that the defendant has done something positive to avoid the commission of an offence.

As a general principle, the case law[2] suggests that a defendant wishing to use this defence must prove that a system was set up to avoid contravention of the statute and that this system was subjected to periodic checks to see that it was working satisfactorily. In the estate agency context, such a 'system' might involve such things as training staff on how to respond to enquiries, ensuring that sales particulars are always checked before printing, and so on. In this connection it should be noted that it is not enough for an employer merely to give instructions and assume that they will be carried out.[3] However, an employer is entitled to delegate responsibility for the firm's compliance system to employees at a senior (supervisory) level, provided that they are not so senior as to rank as the company's 'alter ego'.[4]

One aspect of the 'due diligence' defence is singled out for separate treatment by the 1991 Act, presumably because it is the one which is most likely to arise in practice. This is the question of reliance by the defendant on information given by another person. Section 2(2)

---

1 The burden is clearly on the defendant, and will require proof on the balance of probabilities: *R* v *Carr-Briant* [1943] KB 607.
2 Which is collected in *O'Keefe: The Law relating to Trade Descriptions*, para 250 *et seq*.
3 *Aitchison* v *Reith and Anderson (Dingwall & Tain) Ltd* 1974 SLT 282.
4 See *Tesco Supermarkets Ltd* v *Nattrass* [1972] AC 153.

provides that, in order to rely on this as a defence against liability, the defendant must prove that reliance on the information was reasonable in all the circumstances. It is further explicitly stated that, in considering the 'reasonableness' of reliance, three questions are of particular importance. These are:

(i)   what steps (if any) were taken to check the information;
(ii) what steps (if any) might reasonably have been taken to check it; and
(iii) whether there was any reason to disbelieve the information.

In *Enfield LBC* v *Castles Estate Agents Ltd*,[1] where the defendants were prosecuted for describing as a 'bungalow' a building which did not in fact have planning permission for residential use, they claimed that they had based this description on information received from the vendors of the property and that they had no reason to doubt that information. The Queen's Bench Divisional Court, agreeing with the magistrates, upheld this defence, on the basis that 'bungalow' in question was a substantial brick-built structure, set in its own grounds and with its own driveway leading from the road. However, the court made it clear that, if an agent were to find a 'newly completed and ugly extension tacked on to the side of a house in a conservation area', due diligence would require something more than reliance on the word of the vendor (such as enquiries of the planning authority) before describing the property in any way which asserted that it had planning permission. Interestingly, the court attached little or no weight to the agents' standard procedure, which was to obtain the vendor's written confirmation of advertisements and sale particulars. As Kennedy LJ pointed out, this could hardly amount to verification of information which the vendor had already supplied orally; the crucial question was whether the agent was entitled to rely without corroboration on the vendor's word at all.

Subsections (3) and (4) of section 2 make additional provision for the situation in which a defendant's plea of 'due diligence' amounts in effect to placing the blame on someone else. If the defence involves an allegation that the offence is due to either the 'act or default' of, or 'reliance on information given by' another person, the defendant must serve notice on the prosecution, at least seven days before the trial,

---

1   [1996] 2 EGLR 21.

giving as much information as possible to help identify the person concerned.[1] It has further been held, in cases arising under the Trade Descriptions Act, that it is not sufficient for a defendant merely to name all employees and claim that the offence must have been caused by one of them; the defendant must go on to show that all reasonable steps have been taken to identify the culprit.[2]

Enforcement of the Act

The enforcement provisions of the Property Misdescriptions Act 1991, which are contained in section 3 and the Schedule to the Act, are based on those of the Trade Descriptions Act 1968. Enforcement of the Act is thus the duty of the trading standards departments of local authorities or, in Northern Ireland, of the Department of Economic Development's Trading Standards Branch.[3]

The specific powers of investigation given to trading standards officers resemble those which may be used in investigating criminal offences under the Estate Agents Act 1979, although there are some important differences. Briefly, the powers (all of which are exercisable only by a duly authorised officer of an enforcement authority, only at a reasonable hour and only on production of credentials if demanded) are:

(i)   to enter premises (other than purely residential ones) and to inspect goods, for the purpose of ascertaining whether an offence has been committed[4]

(ii)   on reasonable suspicion of an offence, to inspect books or documents, or a legible reproduction of information stored in other forms (for example, in a computer) and to take copies

(iii) on reasonable belief[5] that they are required as evidence in a prosecution, to seize and detain any books or documents.

---

1   Whether that person can then be prosecuted depends on whether he or she comes within the Act, as a principal or employee of an estate agency or property development business.
2   *McGuire* v *Sittingbourne Co-operative Society* (1976) 140 JP 306.
3   However, all prosecutions in Scotland are initiated by the Procurator Fiscal.
4   Note that, in contrast to the Estate Agents Act, there is no requirement of reasonable suspicion; these powers can be exercised in order to carry out spot checks.
5   This denotes something more than mere suspicion of an offence.

It should be emphasised that the power of entry under this Act does not extend to premises used only as a dwelling, even where an alleged offence concerns a misdescription of that dwelling. Indeed, the power is more restrictive than that contained in the Estate Agents Act, in that, even where a warrant is issued (for example, to justify an entry by force), this cannot extend to purely private premises.

It is a criminal offence to obstruct an authorised officer who is exercising any of the statutory powers mentioned above or, in some circumstances, to fail to give the officer adequate assistance in the exercise of those powers.[1]

---

1   See *Lancashire CC* v   [2007] EWHC 3194 (Admin).

Deposits and Other Payments

6

This chapter is concerned with various sums of money which an estate agent or auctioneer may receive, in the course of marketing property on behalf of a client. In so far as these sums belong to the client, the general law of agency requires the agent to keep them separate from his or her own funds and to keep proper accounts and records of all dealings with the money. These legal obligations are commonly echoed, and sometimes extended,[1] by the members' accounts rules of estate agents' professional bodies.

In the case of deposits paid by a purchaser or potential purchaser, additional issues arise: the statutory rules (under the Estate Agents Act 1979) governing precisely how the money is to be handled; the agent's duty to account for interest; and the extent to which, if the agent defaults, the client may be liable to the purchaser for the loss of the deposit. We shall also note the way in which a landlord client has been held responsible when an estate agent, acting without the landlord's authority or even knowledge, has demanded and received a premium from a residential tenant in circumstances where it is unlawful to do so.

Deposits and deposit-holders

A deposit, the most important type of payment found within the estate agency context, may be of two types: 'contract' or 'pre-contract'. A contract deposit is that sum of money (commonly 10 per cent of the

---

1   For example by applying similar requirements to money which belongs, not just to clients, but to anyone other than the agent.

agreed purchase price) which is paid by the purchaser at the time the contract of sale is made. According to Lord Macnaghten:[1] 'The deposit serves two purposes — if the purchase is carried out it goes against the purchase money — but its primary purpose is this, it is a guarantee that the purchaser means business.' This 'guarantee' may be paid directly to the vendor, but in most cases the contract will provide for it to be handed over to a third party, normally either the auctioneer or the vendor's solicitor.

A pre-contract deposit[2] is a payment, usually of a small amount, made by a potential purchaser before any binding contract is entered into and returnable to that purchaser on demand. The purpose of such a payment is far from clear, since it appears to have no effect whatsoever on the legal position of any of the parties. The purchaser acquires no rights to the property by virtue of making such a payment; the vendor cannot prevent the purchaser from insisting on the return of the money;[3] and the estate agent cannot use this sum as security for a commission claim unless and until it becomes the vendor's money, which will not be before the sale is completed.[4] Furthermore, it appears that, unless the estate agent accepts this money 'as agent for the vendor' with specific authority from the vendor so to do, it must not be paid to the vendor, but must be held at all times on behalf of the purchaser.[5] It was at one time the law that the estate agent was entitled to keep whatever interest might be earned on the money while it was being held,[6] but even this benefit has largely disappeared under the provisions of the Estate Agents Act 1979.[7] As a result, it seems that the only justification for demanding or accepting a pre-contract deposit is a purely practical one, namely, that the estate agent is likely to discover, from a demand for repayment, that a prospective purchaser is no longer interested in acquiring the property.

An estate agent who holds a pre-contract deposit does so as agent for the purchaser, who at all times remains entitled to demand repayment of the money. By contrast, a nominated third party holding a contract deposit may be acting in either of two capacities: as agent for

---

1   *Soper* v *Arnold* (1889) 14 App Cas 429, 435.
2   Sometimes referred to as 'earnest money'.
3   *Chillingworth* v *Esche* [1924] 1 Ch 97.
4   *Skinner* v *The Trustee of the Property of Reed* [1967] Ch 1194.
5   *Rayner* v *Paskell and Cann* (1948) 152 EG 270.
6   *Potters* v *Loppert* [1973] Ch 399.
7   See p 260.

the vendor or as stakeholder, ie a person who 'does not receive the money for either party, he receives it for both; and until the event is known, it is his duty to keep it in his own hands'.[1]

Which of these possibilities applies in any particular case depends primarily on the terms of the contract of sale. Where the contract is silent, it is presumed that the vendor's solicitor holds the deposit as agent for the vendor,[2] although this presumption is in no way conclusive. In *Wiggins* v *Lord*,[3] for example, where the contract of sale provided for the payment of the deposit to certain named persons, who were in fact the vendor's solicitors though not described as such, it was held that they received the money as stakeholders.

An auctioneer, on the other hand is presumed to receive a deposit as stakeholder, and this presumption is a strong one. Thus, in *Furtado* v *Lumley*[4] an auctioneer was held to be a stakeholder, notwithstanding that the auctioneer had signed the conditions of sale 'as agent' and that the conditions made express provision for the vendor to return the deposit in certain circumstances. Similarly in *Edwards* v *Hodding*,[5] where the auctioneer was also solicitor to the vendors, it was held that the auctioneer must be treated as a stakeholder of the deposit. Given this strong presumption, it is surprising, to say the least, that to find suggestions from the Court of Appeal that an estate agent would be presumed to hold a contract deposit as agent for the vendor.[6]

Vendor's responsibility for lost deposits

Where money paid by a purchaser to the vendor's agent subsequently goes astray (whether or not through dishonesty on the part of the agent), the question to be answered is whether the purchaser remains indebted to the vendor, or whether the payment to the agent has effectively discharged his or her debt. As is so often the case in agency transactions, the answer to this question depends on the extent of the agent's authority: if the agent was expressly or impliedly authorised by the vendor to receive payment, the purchaser's debt is discharged; if not, the purchaser can be forced to pay again.

---

1   *Harington* v *Hoggart* (1830) 1 B & Ad 577, 586, *per* Lord Tenterden CJ.
2   *Edgell* v *Day* (1965) LR 1 CP 80.
3   (1841) 4 Beav 30. See also *Wolf* v *Hosier & Dickinson* [1981] Com LR 89.
4   (1890) 6 TLR 168.
5   (1814) 5 Taunt 815.
6   *Ojelay* v *Neosale Ltd* [1987] 2 EGLR 167.

Contract deposits

Agent's authority to receive payment

In the absence of express authority from the vendor to accept a deposit, the question is whether such authority can be implied from the agent's role in bringing about the contract of sale. As to this, two propositions may be confidently stated. First, an estate agent who has no authority to make a binding contract of sale on behalf of the client[1] equally has no implied authority to receive payment of the deposit which is paid on exchange of contracts. Second, by contrast, where a contract of sale specifically requires a deposit to be paid to an estate agent or (more commonly) an auctioneer, that agent will thereby be authorised to receive it on the vendor's behalf.

The difficult question is whether authority for an agent to conclude a sale impliedly carries with it authority to receive a deposit. The courts at one time insisted that it did so; as long ago as 1698, Lord Holt CJ was reported as saying:[2] 'He who has power to sell, has power to receive the money.' Similar sentiments were expressed by Lord Tenterden CJ in *Capel* v *Thornton*:[3] 'If he, as their agent, had authority to sell goods, so had he (in the absence of advice to the contrary), an implied authority to receive the proceeds of such sale.' In subsequent cases, however, the courts treated these early statements as applying only to certain classes of agents, such as factors, and stated the general rule to be quite the reverse. Thus, in *Drakeford* v *Piercey*,[4] Lush J said: 'that an agent authorised to sell has as a necessary legal consequence authority to receive payment is a proposition utterly untenable and contrary to authority'.

None of the cases mentioned in the previous paragraph dealt specifically with an auctioneer, probably because the issue at auction is almost invariably settled by the conditions of sale. However, the case of *Mynn* v *Joliffe*[5] lends some support to the view that an auctioneer in this respect has no greater authority the general run of agents. The claimant there, who had bought land at an auction and had paid a deposit to the auctioneer, sued the vendor to recover this money when

---

1   This is almost invariably the position: see Chapter 4.
2   Anon (1698) 12 Mood 230.
3   (1828) 3 C&P 352. See also *Howard* v *Chapman* (1831) 4 C&P 508.
4   (1866) 7 B&S 515, 522. Followed in *Butwick* v *Grant* [1924] 2 KB 483.
5   (1834) 1 Moo & R 326.

the vendor failed to make a good title to the property. Counsel for the vendor objected that no evidence had been led to establish that the auctioneer had been expressly authorised to receive the deposit, and claimed that, in the absence of express authority, none could be implied. To this Littledale J replied: 'I think that an agent employed to sell has no authority, as such, to receive payment.'

As far as estate agents are concerned, the question has never arisen in the English courts, presumably because estate agents are so seldom authorised to conclude a contract of sale. However, it has been held by a New Zealand court that an estate agent who is instructed to make a contract for the sale of property, on terms which require the payment of a deposit, has in consequence implied authority to receive that deposit on the vendor's behalf.[1]

Method of payment

As already stated, the authority of an auctioneer almost invariably derives from the conditions of sale, which commonly provide for payment to the auctioneer of the full purchase price of goods and the deposit on a sale of land. The fact that the auctioneer is so authorised, however, is not the end of the matter; in the absence of any express provision to the contrary, it is assumed that payment will be in cash, and a principal is not bound by an agent's acceptance of any alternative form of payment. As Lord Loughborough CJ said:[2]

> In the common course of auctions, there is no delivery without actual payment; if it be otherwise, the auctioneer gives credit to the vendee, entirely at his own risk.

In *Brown* v *Staton*[3] the purchaser of goods told the auctioneer that the vendor owed him money, and the auctioneer allowed him to set this off against the price of the goods. The auctioneer was held liable to account to the vendor for the full sale price of the goods, and Bayley J said: 'It cannot be contended that an agent has duly accounted unless he pays over the full price.'

Thus an agent employed to sell goods is not authorised to pay off the principal's debts; nor, it seems, is such an agent entitled to take

---

1   *Boote* v *R T Shiels & Co* [1978] 1 NZLR 445.
2   *Williams* v *Millington* (1788) HBl 81, 84.
3   (1816) 2 Chit 353.

payment in kind.[1] Most of the cases, however, in which the scope of an agent's authority has been considered have concerned payments made by some form of negotiable instrument, and the general rule has emerged that an agent has no implied authority to accept a cheque, bill of exchange, or IOU, at least as payment of the full price.[2] In *Earl of Ferrers* v *Robins*[3] the defendant was instructed to sell furniture belonging to the claimant by auction for ready money. He accepted a bill of exchange drawn by the purchaser on a third party, which the claimant subsequently rejected. The defendant, when sued for breach of his instructions, was forced to admit, not only this specific breach, but also more generally that the vendor was quite entitled to refuse the bill of exchange in payment.

In *Sykes* v *Giles*,[4] likewise, the actual decision turned on the express instructions given to the auctioneer. In that case the conditions governing a sale of growing beech timber laid down that the purchaser of each lot should pay a deposit of 10 per cent to the auctioneer. The remainder of the price was to be paid by a stipulated date and security for the due performance of this obligation was to be given to the vendor. The defendant purchased a particular lot and paid a deposit. Later the auctioneers, who were in financial difficulties, induced the purchaser to give them a bill of exchange for the remainder of the price, which they then cashed. The vendor sued the purchaser for payment and succeeded. In the opinion of the court, the conditions of sale made it clear to all concerned that the balance of the purchase price was to be paid directly to the vendor, and not to an agent. 'But even if the auctioneer had had authority to receive the remainder of the purchase-money, he had no authority to receive it in this way, by means of a bill of exchange. Cash payment was intended, and not a bill of exchange.'[5]

In *Williams* v *Evans*[6] no express conditions obscured the application of the general rule. In that case a vendor, who suspected the auctioneer's solvency, told a purchaser not to pay any money to the auctioneer. The purchaser, however, had already given the auctioneer

---

1   In *Howard* v *Chapman* (1831) 4 C & P 508 the defendant, who had sought to discharge a debt by giving the claimant's agent a quantity of horse-hair, was held still liable to pay the claimant.

2   As to the payment of a deposit by cheque, see pp 239–240.

3   1835) 2 Cr M & R 152.

4   (1839) 5 M & W 645.

5   At p 652, *per* Parke B.

6   (1866) LR 1 QB 352.

a bill of exchange for part of the price and this was duly honoured on being presented two days later. The question for decision was therefore whether, when the auctioneer accepted the bill of exchange, he had authority to do so. Blackburn J, delivering the judgment of the Court of Queen's Bench, had no doubts:[1]

> I think, on the authority of *Sykes* v *Giles*, payment by a bill of exchange to an auctioneer is not a valid payment for the purpose of discharging the debtor.

The purchaser therefore remained liable to the vendor for the price.

The Irish case of *Hodgens* v *Keon*[2] illustrates the position of all three parties where a purchaser gives the auctioneer a bill of exchange. The defendant bid for leasehold property at an auction and, on being declared purchaser, claimed that he was bidding as agent for someone else. He gave the auctioneer an IOU for the deposit and in due course the auctioneer sued upon this. The defendant argued that, even if he paid the auctioneer, he would still remain liable to the vendor, who had not authorised his agent to accept payment in this way. The defendant was nevertheless held liable. Indeed, as emerged from the judgment of Palles CB, this was only one instance where one person might be liable to two, or might be entitled to sue two. It was said that if the vendor had authorised the auctioneer to receive an IOU, the vendor's sole action for the money would lie against the auctioneer, who would in turn be able to sue the purchaser. If, however, the auctioneer had no such authority, both auctioneer and purchaser would be liable to the vendor, and both vendor and auctioneer could sue the purchaser.

It should be noted that where a payment is made by cheque, and the cheque is duly honoured on presentation, it is the equivalent of a payment in cash.[3] However, it is only so treated from the moment at which it is honoured; thus, as we saw in *Williams* v *Evans*,[4] if the agent's authority to receive payment at all is revoked during the currency of a bill of exchange, and there was no authority at the time to take a bill of exchange, the purchaser's debt is not discharged.

The general rule, that an auctioneer may only accept cash, appears to have been made subject to one qualification by the decision of the

---

1   At p 355.
2   [1894] 2 IR 657.
3   *Bridges* v *Garrett* (1870) LR 5 CP 451.
4   (1866) LR 1 QB 352.

Court of Appeal in *Farrer* v *Lacy, Hartland & Co.*[1] Property there was auctioned on behalf of a mortgagee and the auctioneer, with the concurrence of the mortgagee, took a cheque for the deposit. When this cheque was dishonoured on presentation, the mortgagor claimed that the auctioneer's negligence was such as to deprive the mortgagee of the right to claim the costs of the abortive sale. It was held that the auctioneer was not guilty of negligence, as he had merely acted in accordance with the common practice at auctions. As Bowen LJ said:[2]

> Can it be said that 99 out of every 100 sales are conducted unreasonably because cheques are taken by way of payment of deposit? On the contrary, it was, I think a reasonable thing ... to do.

Furthermore, as Baggallay LJ pointed out:[3] 'Persons cannot be expected to come to sales with large sums of money in their pocket.'

It appears, then, that an auctioneer may accept a cheque for the deposit, provided always, of course, that it is reasonable to do so in the circumstances; to accept a cheque despite having reason to suspect the creditworthiness of the purchaser would lay the auctioneer himself open to a charge of negligence by the client.

Nevertheless, the fact that an auctioneer has authority to accept a cheque as payment of a deposit does not entitle the purchaser to insist on paying in this way. In *Johnston* v *Boyes*[4] the claimant, a married woman, instructed her husband to bid for a public-house, and he did so successfully. The vendor, however, recognised the husband as a man of little resources and refused to accept a cheque, whereupon the auctioneer resold the property. The claimant claimed that her husband had been wrongfully prevented from signing the contract of sale; it was held, however, that a purchaser's basic obligation is to pay a deposit in cash and, since this had not been done, no liability attached to the vendor.

Default by stakeholder

Where a stakeholder absconds with the stake or becomes insolvent, one of two innocent parties must suffer the loss. It might be expected

---

1   (1885) 31 ChD 42.
2   At p 48.
3   At p 46.
4   [1899] 2 Ch 73.

that this would simply fall upon the person who ultimately becomes entitled to receive the money from the stakeholder. That is indeed the position taken by the law in cases of true 'stakeholding', where the parties to a wager jointly nominate a third party to hold the bet.

Where, however, the stakeholder is the nominee of one party alone (as in the case of an auctioneer) it appears that the nominating party is treated as taking responsibility for the stakeholder's honesty and solvency. In *Fenton* v *Browne*[1] a sale of land took place by private treaty, though under conditions of sale drafted for an auction. A deposit was paid to a stakeholder and, when disputes arose, the purchaser asked for it to be paid into court. The vendor refused to agree to this proposal; consequently, when the agent absconded with the deposit, the vendor was held to have taken the risk upon himself. The decision, then, turned upon the special circumstances of the case; however, Sir William Grant MR took the opportunity to comment upon the general situation:[2]

> Upon a sale by auction the vendor determines who is to receive the deposit. The auctioneer is not a stakeholder of the purchaser: at least not of his choice. If he were a stakeholder for both parties, either would have a right to propose to change the stakeholder; and the party refusing takes upon himself the risk.

This *dictum* was quoted with approval in *Smith* v *Jackson*,[3] where a deposit was paid, according to the conditions of sale, as part of the purchase money. It was held that it must be treated, while in the auctioneer's hands, as the vendor's money, so that the purchaser was entitled to credit for the interest which it earned. The case of *Annesley* v *Muggridge*[4] was likewise not directly in point, for it dealt with an auctioneer's claim to deduct charges and expenses from a deposit before paying it into court. In deciding that the auctioneer could lawfully do this, on the ground that the deposit must be treated as the vendor's money, Sir Thomas Plumer V-C said:[5]

> Pending the dispute as to the title, all the risk respecting the deposit rests with the vendor: for though the auctioneer is, to a certain degree, a

---

1   (1807) 14 Ves 144.
2   At p 150.
3   (1816) 1 Madd 618.
4   (1816) 1 Madd 593.
5   At p 596.

stakeholder for vendor and vendee, yet so far as respects any risk as to the deposit, the auctioneer is considered as the agent only of the vendor.

The proposition supported by these various *dicta* was placed beyond doubt by the case of *Rowe* v *May*,[1] where a mortgagee put property up for auction in the exercise of a power of sale. The property failed to sell, whereupon the mortgagor instructed the auctioneer to sell it privately on the same conditions. The auctioneer having done so, and having taken a deposit, the sale was adopted by the mortgagee. When it was later discovered that the auctioneer was insolvent and the deposit lost, the mortgagee was held liable to complete the sale upon the purchaser paying only the balance of the price. Sir John Romilly MR said:[2]

> Where a purchaser pays a deposit on his purchase-money to the auctioneer, and it is lost, on whom does the loss fall? If the matter goes off, because the vendor cannot make a good title, it is the vendor's duty to repay the deposit, and the loss occasioned by the non-completion; and in case an action were brought against him for breach of the contract, the amount of deposit not repaid would be part of his loss, and the purchaser would be entitled to add it to the damages; so if the contract be completed, and the deposit cannot be recovered from the auctioneer, who for this purpose is the agent of the vendor, it will be the vendor's loss, and not that of the purchaser.

Not everyone would agree that it is morally justifiable to make the vendor responsible for the auctioneer's default. Moreover, the authorities cited above, on which this principle is based, are all of some age. However, such recent judicial reference as there has been to the point suggests that it remains valid.[3]

Pre-contract deposits

A particular problem which has been caused by pre-contract deposits, and the one with which we are presently concerned, arises when an

---

1   (1854) 18 Beav 613.
2   At p 616.
3   See *Burt* v *Claude Cousins & Co Ltd* [1971] 2 QB 426, 450, *per* Sachs LJ; *Barrington* v *Lee* [1972] 1 QB 326, 335, *per* Lord Denning MR; *Sorrell* v *Finch* [1977] AC 728, 754, *per* Lord Russell of Killowen.

estate agent absconds with the money. Naturally, a vendor who has actually authorised the agent to receive this sum will be liable to the purchaser for its repayment;[1] in the absence of express authority from the vendor, however, the question is whether such authority is to be implied. In a series of cases which began with the decision of the Court of Appeal in *Ryan* v *Pilkington*,[2] the English courts held that such implied authority did indeed exist. A vendor was thus responsible to the purchaser for money, notwithstanding that the vendor had not asked the agent to collect it, might indeed know nothing about it and certainly had no legal entitlement to it. The legal justification for this extraordinary result was, according to Morris LJ in *Ryan* v *Pilkington*, that the collection of earnest money is reasonably incidental to the function of finding a purchaser and thus within an estate agent's implied authority. The moral justification was summed up by Megaw LJ in the later case of *Burt* v *Claude Cousins & Co Ltd*:[3]

> It is the vendor who has chosen the estate agent, who has clothed him with the capacity of agent, and who has enabled him to ask for and receive a deposit in connection with the business to which the agency relates.

This self-perpetuating line of authority, which appeared both legally and morally suspect, was finally overturned by the House of Lords in the case of *Sorrell* v *Finch*.[4] In the opinion of their lordships, the argument that the vendor should be liable on the ground of having chosen the estate agent was 'too simplistic'; more important was the point that a pre-contract deposit remains at all times the property of the purchaser. As Lord Edmund-Davies pointed out:[5]

> The just solution of the problem of which of two innocent parties should suffer should depend very largely (and possibly conclusively) upon what rights could have been asserted by each of them in respect of the money in the agent's hands at all material times.

---

1   *Duke of Norfolk* v *Worthy* (1808) 1 Camp 337.
2   [1959] 1 All ER 689. Of the others, the most important are *Goding* v *Frazer* [1966] 3 All ER 234; *Hassan* v *Varol Agencies Ltd* (1969) 212 EG 39; *Maloney* v *Hardy and Moorshead* (1970) 216 EG 1582; *Burt* v *Claude Cousins & Co Ltd* [1971] 2 QB 426; *Barrington* v *Lee* [1972] 1 QB 326.
3   [1971] 2 QB 426, 452.
4   [1977] AC 728.
5   *Ibid*, at p 748.

The illogicality and injustice of making a vendor responsible for money to which he or she had no claim was equally important in the eyes of Lord Russell of Killowen:[1]

> I find it impossible to accept a theory of authority from a vendor to the estate agent to accept a deposit *as agent for the vendor* on terms that so long as all remains in negotiation the estate agent must on no account pay the deposit to the vendor or to the vendor's order, but on the contrary must on demand of the purchaser pay it back to the purchaser without reference to the vendor ... That appears to me with all respect to be juridically unacceptable.

As a result of the decision in *Sorrell* v *Finch*, it is now clearly established that, unless an estate agent is expressly authorised to receive a pre-contract deposit on behalf of a client, it is the purchaser alone who takes the risk that the agent will default in repayment. Moreover, this applies irrespective of whether the estate agent purports to receive the money 'as agent', 'as stakeholder' or without specifying any particular capacity. This change from the previous unjust and illogical rule is greatly to be welcomed, but it should not be thought that the problem is thereby completely solved. A lost deposit, notwithstanding *Sorrell* v *Finch*, remains lost; only the identity of the loser is changed. While the major professional bodies to which estate agents belong provide a measure of protection by means of Accounts Rules and Compensation Schemes, these do not cover the kind of agent who appeared in *Sorrell* v *Finch* itself:

> The estate agent was, so far as is known, a man with no professional qualifications and a member of no professional organisation. He was, unknown to either the vendor or to any of the possible purchasers concerned, a dishonest rogue and an undischarged bankrupt.[2]

It is greatly to be regretted, therefore, that the Government has so far declined to implement the full range of deposit-protecting powers which are available to it under the provisions of the Estate Agents Act 1979.[3]

---

1   [1977] AC 728, 753.
2   [1977] AC 728, 752, *per* Lord Russell of Killowen.
3   See p 251.

Agent's personal responsibility

Where a person holding a deposit does so as agent for the vendor, either because the contract so provides or because that is the inference drawn from the circumstances by the court, payment to the agent is deemed at the same moment to be payment to the principal. As a result, a purchaser who becomes entitled to the return of the money can recover it from the vendor, whether or not the latter has received it from the agent.[1] The purchaser cannot claim against the agent in such circumstances, even where the deposit is still in the agent's possession.[2]

A person to whom money is paid as stakeholder, on the other hand, undertakes personal responsibility for its safe keeping. Such a person must await the event on which one party or the other becomes entitled to the money; if, before that event occurs, the money is paid to one party without the express consent of the other, the stakeholder is liable. In *Burrough* v *Skinner*[3] the claimant, who had bought an interest in land at an auction, paid a deposit of £50 to the defendant auctioneer. The sale fell through and the auctioneer was held liable to return the deposit. It was said of the auctioneer that: 'he was a stakeholder, a mere depositary of the £50 and ought not to have parted with it, till such time as the sale should be finished and completed, and it should appear in the event to whom it belonged.'

Although the report of that case does not make it clear, it may be surmised that the auctioneer had paid over the deposit to the vendor. Certainly it has been held in later cases that such a payment by the auctioneer is a breach of his duty as stakeholder, and *Burrough* v *Skinner* has been cited as the basis of the rule. A number of the cases, however, can be explained on other grounds. Thus in *Edwards* v *Hodding*,[4] where the auctioneer was also the vendor's solicitor, he clearly knew before paying over the deposit that the vendor could not make out a good title to the property. Similarly in *Gray* v *Gutteridge*,[5] where *Burrough* v *Skinner* was again relied upon, it appeared that the auctioneer had, immediately after the sale, entered into a personal contract with the purchaser to see that the sale was completed.

---

1   *Duke of Norfolk* v *Worthy* (1808) 1 Camp 337.
2   *Bamford* v *Shuttleworth* (1840) 11 A & E 926; *Ellis* v *Goulton* [1893] 1 QB 350.
3   (1770) 5 Burr 263.
4   (1814) 5 Taunt 815.
5   (1828) 1 Man & R 614.

Whether or not these two cases support in any binding fashion the proposition that an auctioneer may not pay the deposit to the vendor until the latter is entitled to it under the contract is uncertain. No such uncertainty arises, however, in respect of the decision of Stephen J in *Furtado* v *Lumley*.[1] In that case the defendant auctioneer paid over to the vendor part of a deposit received on the sale of his property. The sale having fallen through, the defendant returned the remainder of the deposit money to the purchaser, who sued the vendor for the balance. When the vendor proved unable to pay this sum, the purchaser was held entitled to recover it from the defendant.

Precisely what is involved in a stakeholder's obligations has been considered in cases concerning solicitors who received deposits expressly 'as stakeholders'. In *Hastingwood Property Ltd* v *Saunders Bearman Anselm*[2] it was held that a stakeholder who is convinced that one of the parties has become entitled to the deposit is both bound and entitled to pay it over to that party. The other party cannot prevent this payment by injunction, although a stakeholder whose decision turns out to be wrong will of course incur personal liability for this. On the other hand, where both parties instruct the stakeholder to deal with the deposit in a particular way, their instructions must be obeyed; the stakeholder cannot insist on retaining the money, for example because of some claim which the stakeholder is making against one of the parties.[3]

Estate agents and deposits — the statutory regime

The intention which underlies sections 12 to 17 of the 1979 Act, and the Estate Agents (Accounts) Regulations 1981, is that estate agents should deal properly with clients' money and that, in the event of an agent's default, the public should be protected as far as possible against any resulting loss. In considering this matter, it will be convenient to look first at the statutory definition of 'clients' money', before turning to the specific obligations which the Act imposes in respect of such money.

---

1   (1890) 6 TLR 168.
2   [1990] 3 All ER 107.
3   *Rockeagle Ltd* v *Alsop Wilkinson* [1991] 4 All ER 659.

'Clients' money' in the Estate Agents Act 1979

For the purposes of the Estate Agents Act, 'clients' money' bears a restricted meaning, as is apparent from the wording of section 12(1):

> In this Act 'clients' money', in relation to a person engaged in estate agency work, means any money received by him in the course of that work which is a contract or pre-contract deposit —
>
> (a) in respect of the acquisition of an interest in land in the United Kingdom, or
>
> (b) in respect of a connected contract,
>
> whether that money is held or received by him as agent, bailee, stakeholder or in any other capacity.

Of the various sums of money which may be handed over in the course of estate agency work, therefore, the Act controls only the two types of deposit on sales of land and contracts connected to and conditional upon sales of land, such as sales of curtains and carpets. Section 12(2) provides:

> In this Act 'contract deposit' means any sum paid by a purchaser —
>
> (a) which in whole or in part is, or is intended to form part of, the consideration for acquiring [an interest in land in the United Kingdom] or for a connected contract; and
>
> (b) which is paid by him at or after the time at which he acquires the interest or enters into an enforceable contract to acquire it.

Section 12(3) applies a similar definition to 'pre-contract deposit' except that in this case, of course, the sum is paid before the interest is acquired or the contract made.

In view of the obligations which attach to 'clients' money' and the severe consequences of failure to meet them, the definitions in section 12 are of fundamental importance, and several points are worthy of emphasis. First, since the definition of 'clients' money' includes both contract and pre-contract deposits, the question into which of these categories a particular sum falls is one of purely academic interest. Second, it matters not whether the money is received by the estate agent as agent, as stakeholder or in any other capacity;[1] the Act makes

---

1 The word 'bailee' seems inappropriate in this situation, as this usually refers to the deposit of goods rather than money.

it clear that the same rules apply. Thirdly, in cases where a sum of money is paid to an estate agent partly as a deposit and partly for other purposes (eg for transmission to a building society or an insurance company) then, if it cannot conveniently be split, the whole sum is 'clients' money' and must therefore pass through the 'client account'.

Clients' money as trust property

The problems caused in the past by estate agents who defaulted with deposit-money were due in no small measure to the common law rule that the legal claim of a person entitled to a deposit against the deposit-holder was a purely personal one. Hence, if the deposit-holders' assets were insufficient to pay off general creditors, the deposit was in no way 'earmarked' for the person entitled to it. Nor, if the deposit-holder had wrongfully used the money, could it be traced. In an attempt to alleviate the hardship caused by this rule, section 13(1) provides that where clients' money, as defined above, is received by any person in the course of estate agency work, that money

(a) is held by him on trust for the person who is entitled to call for it to be paid over to him or to be paid on his direction or to have it otherwise credited to him, or

(b) if it is received by him as stakeholder, is held by him on trust for the person who may become so entitled on the occurrence of the event against which the money is held.[1]

This statutory creation of a trust in respect of clients' money has a number of important effects. First, it means that, if an estate agent becomes insolvent, any money which is in that agent's 'client account' cannot be used to pay off general creditors; it is 'property held by the bankrupt on trust for any other person'[2] and is therefore payable only to those persons entitled to the deposits.[3] Second, if the estate agent

---

1 If the estate agency work is done in Scotland (no matter where in the United Kingdom the property is located) the money is held 'as agent' rather than on trust (s 13(2)); the legal effect, however, is the same.

2 Insolvency Act 1985, s 130(3).

3 If there is not enough money in the client account to meet all claims, the rule in *Clayton*'s case (1816) 1 Mer 572 applies. It is presumed that withdrawals from the account are made in the same order as payments in; hence, 'first in, first out'.

has wrongfully used money impressed with a trust, for example by paying it into his or her own bank account or buying goods with it, the person entitled to the money may be able to 'trace' it, even into a different form and into the hands of a third party (though not of a bona fide purchaser for value without notice).[1] Less importantly, the existence of a trust brings into operation various supervisory powers of the High Court, notably the power to appoint a new trustee where necessary, such as where the sole principal of an estate agency practice dies or becomes insane.[2]

The designation of clients' money as a trust fund would normally lead to a number of other consequences under the general law of trusts. Those responsible for the drafting of the Estate Agents Act, however, adopted a more specific approach; according to section 13(3):

> The provisions of sections 14 and 15 below as to the investment of clients' money, the keeping of accounts and records and accounting for interest shall have effect in place of the corresponding duties which would be owed by a person holding clients' money as trustee, or in Scotland as agent, under the general law.

Further, it is provided by section 13(5) ('for the avoidance of doubt') that, where an estate agent has a lien over clients' money, this does not affect the trust; equally, the trust does not affect the lien.[3]

The statutory trust created by section 13(1) is expressed to be for the benefit of the person who is entitled to demand the clients' money from the estate agent or, in certain circumstances where nobody is so entitled, to the person who may become entitled. In the vast majority

---

1   The equitable rules on 'tracing' are complex, depending mainly upon the status of the ultimate recipient of the money and also upon whether the money remains sufficiently 'identifiable'. Readers are referred to any of the major works on Equity.
2   In the particular case where an order made by the Director-General of Fair Trading prohibits an estate agent from holding clients' money, it is provided that the order itself may appoint a substitute trustee and make provision for his expenses and remuneration: s 13(4).
3   Although it has never been specifically decided whether or not an estate agent has a lien on clients' money for remuneration and expenses, it seems in principle (by analogy with the case of an auctioneer: *Webb* v *Smith* (1885) 30 ChD 192) that there should be one on money which belongs to the client. In practice, this will only apply to a contract deposit which is still in the agent's hands after the sale is completed or (possibly) to a contract deposit which has been received 'as agent' with the authority of the vendor.

of cases, this will be one of the parties to an actual or proposed sale; which one it is will depend upon the type of deposit in question and the capacity in which it is received and held by the estate agent. In practical terms, it appears that the 'person entitled' to a pre-contract deposit is the prospective purchaser who has paid it, irrespective of whether the estate agent purports to receive that deposit 'as agent for the vendor', or 'as stakeholder', or without any designation.[1] The only exceptions to this general principle are where the vendor has expressly authorised the agent to receive a pre-contract deposit as agent, or where the agent purports (without authority) so to receive it and the vendor subsequently ratifies what the agent has done. If, while an estate agent is holding a pre-contract deposit, contracts are exchanged between vendor and purchaser, the status of that deposit depends upon the terms of the contract. It may well be that the money is expressed to become part of the contract deposit; if the contract is silent, however, the rules outlined above will continue to apply.

In relation to a contract deposit, an estate agent may receive the money either 'as agent for the vendor' or 'as stakeholder'. In the former case (assuming that the agent is acting with the vendor's authority), it is the vendor who is the 'person entitled' for the purposes of section 13.[2] In the latter case, neither party is immediately entitled to demand the money; which of them subsequently becomes entitled depends upon whether the sale is successfully completed (the vendor) or one of the parties wrongfully fails to complete (the innocent party). Perhaps because estate agents so seldom hold contract deposits, the English courts have never been called upon to decide the capacity in which they are presumed to do so in the absence of any evidence to the contrary. Given that an auctioneer is presumed to be a stakeholder,[3] it would seem appropriate to regard an estate agent in a similar light; however, *dicta* from the Court of Appeal suggest that the presumption is in favour of holding the deposit 'as agent'.[4]

---

1   *Sorrell* v *Finch* [1977] AC 728: p 243.
2   *Ellis* v *Goulton* [1893] 1 QB 350.
3   *Furtado* v *Lumley* (1890) 6 TLR 168. By contrast, the vendor's solicitor is presumed to hold a contract deposit as agent for his client: *Edgell* v *Day* (1865) LR 1 CP 80. See p 235.
4   *Ojelay* v *Neosale Ltd* [1987] 2 EGLR 167.

Restrictions on deposit-taking

There are four situations in which the Act either curtails or removes altogether the power of an estate agent to accept a deposit. Of these, potentially the most important is probably section 16(1), which provides:

> Subject to the provisions of this section, a person may not accept clients' money in the course of estate agency work unless there are in force authorised arrangements under which, in the event of his failing to account for such money to the person entitled to it, his liability will be made good by another.[1]

Surprisingly, perhaps, this 'compulsory bonding' provision has not yet been brought into force. Even more surprisingly, if it is implemented, it will not require all who engage in estate agency work to be bonded, but only those who accept clients' money. However, in an effort to ensure that members of the public realise when they are dealing with an unbonded agent, section 16(4) makes it a criminal offence for any person who carries on estate agency work to describe himself or herself as an 'estate agent', or to indicate in any way that he or she is in business 'to act as a broker in the acquisition or disposal of interests in land', unless the relevant information[2] is displayed both at the agent's place of business and in all documents, such as advertisements, notices and the like, which may induce people to use the agent's services.

The details of what constitutes 'authorised arrangements' are not set out in the Act; they are to be filled in by subsequent regulations, which may for example specify the circumstances in which the liability of the insurers may be excluded,[3] and provide that any 'ceiling' on that liability in an individual case shall not be less than a specified amount.[4] It is, of course, impossible to predict with certainty the content of such regulations, but it seems safe to assume that the insurance required would have to provide total cover, not only for the agent's deliberate misappropriation of clients' money, but also for the careless mixing of such money with the agent's own funds, which results in its loss on the

---

1   The person entitled to the money will not of course be a party to these 'authorised arrangements', but may nonetheless make a direct claim against the insurers: s 16(3).

2   The content of which will be specified in regulations.

3   S 16(2)(c).

4   S 16(2)(d).

agent's bankruptcy. It also appears that the insurance schemes presently operated by the major professional bodies would be designated as satisfying the requirements of the statute.

Where estate agency work is carried on by such institutions as banks, building societies or insurance companies, whose funds are adequately protected by other legislation, they may be exempted from the requirements of section 16, either under the regulations or on application to the Director-General of Fair Trading under section 17. Subject to such exemption, however, these provisions are regarded as very important; any breach is a criminal offence.

The second situation in which an estate agent may not hold clients' money is where the agent has a personal interest in the property concerned. Section 21(4) provides:

> An estate agent may not seek or receive a contract or pre-contract deposit in respect of the acquisition or proposed acquisition of —
>
> (a) a personal interest of his in land in the United Kingdom; or
> (b) any other interest in any such land in which he has a personal interest.

The odd thing about this provision is that it appears not to apply to the sale of an agent own property. Such a case would not normally fall within the Act at all, because section 1 requires the agent to be acting 'pursuant to instructions', and one can hardly 'instruct' oneself to act. Furthermore, the extension by section 21(3) to private deals applies only to the duty of disclosure imposed by subsections (1) and (2). If this analysis is correct, the ban on seeking or receiving deposits is limited to cases where the estate agent has a 'personal interest' in land but is nevertheless selling it as agent for a client.

It should be noted that, even where an estate agent accepts a contract deposit in breach of this provision, there is no statutory obligation on the agent to refund it.[1] Section 21(6) provides that no breach of the section shall give rise to any form of legal action, either civil or criminal.

The remaining two restrictions on deposit-taking both relate solely to pre-contract deposits. Where estate agency work is done in Scotland (no matter where in the United Kingdom the land itself may be situated), such deposits are banned altogether; section 20 provides that no pre-contract deposit may be sought or received. Where estate

---

1 A pre-contract deposit must in any case be returned on demand.

agency work is carried out in England, Wales or Northern Ireland, there is no such absolute prohibition, but section 19[1] enables regulations to be made which will limit the size of the sums involved 'and such a limit may be so prescribed either as a specific amount or as a percentage or fraction of a price or other amount determined in any particular case in accordance with the regulations'.

Dealing with deposits

When an estate agent receives clients' money in circumstances where this is permitted, the Act leaves no doubt as to what should be done with it. Section 14(1) provides:

> Subject to such provision as may be made by accounts regulations, every person who receives clients' money in the course of estate agency work shall, without delay,[2] pay the money into a client account maintained by him or by a person in whose employment he is.

Failure to comply with this obligation is a criminal offence carrying a fine of not more than £500.

The definition of a 'client account', and the detailed rules which govern its operation, are to be found in section 14 of the Act and the Estate Agents (Accounts) Regulations 1981. According to section 14(2), a client account is a current or deposit account with an authorised institution, which is in the name of a person who is or has been engaged in estate agency work and which contains in its title the word 'client'. The 'authorised institutions', which are listed in the Schedule to the Regulations, are, broadly speaking, banks (including trustee savings banks), building societies and the Post Office. It is further provided by regulation 3 that, should any of these institutions themselves engage in estate agency work, they and their employees are exempt from the requirements relating to client accounts.[3]

Regulations 4 and 5 govern the precise way in which a client account is to operate. The basic objective is to achieve the rigid separation of 'clients' money', as defined by the Act, from all other

---

1   Which has not yet been brought into force.
2   Unfortunately perhaps, in view of the severe consequences of any breach of this provision, 'without delay' is not further defined in the Act.
3   This is on the basis that depositors are given sufficient protection by other statutory codes which govern such institutions.

money in the estate agent's hands; to this end, while clients' money must always be paid into a client account without delay, there are only two other occasions on which other money may lawfully be paid into the same account. These are, first, where the institution in question requires a minimum sum to open or maintain the account (in which case it can and should be withdrawn when it is no longer required for that purpose) and, secondly, to restore in whole or in part any money wrongfully paid out of the account.[1] The regulations also clarify the position of money which is paid to an estate agent for more than one reason. In such circumstances,[2] the whole sum is deemed to be 'clients' money' but, in so far as it is practicable to split that part which may be strictly described as a 'deposit' from the rest, only the former may be paid into the client account. If it is not practicable to split it, then the entire sum must be paid into the account and the surplus removed.

The circumstances in which money may lawfully be paid out of the account (apart of the removal of anything which should never have been paid into it) are:

- where money is paid to the person entitled to it[3]
- in payment of the estate agent's remuneration or expenses, with the agreement of the person for whom the money is held
- in the lawful exercise of a lien[4] and
- where money is transferred to another client account.[5]

It follows from these rules that, strictly speaking, a client account should neither be credited with interest earned on the money in it, nor debited with any bank charges. The agent must arrange for these payments in and out to be allocated to other accounts.

Accounts and records

In an effort to ensure that there is a means of checking compliance by an estate agent with his duty to open a client account and to operate it

---

1 This provision might seem odd but, without it, every wrongful payment out of a client account (which constitutes a criminal offence) would automatically be followed by another offence when the money was returned.
2 See p 248.
3 As to who is the 'person entitled', see p 249–250.
4 See p 152.
5 In which case the records kept must enable its movements to be clearly traced.

properly, section 14 and the Accounts Regulations lay down details of the actual accounts and records which are to be kept.[1] The duty to keep these accounts is imposed by regulation 6 upon 'any person who receives clients' money in the course of estate agency work', except an employee who always pays such money without delay into the employer's client account. It is, however, provided that the obligation (which involves keeping records for six years after the end of the accounting period to which they relate) may be handed over by one person to another when the former ceases to be engaged in estate agency work.

As to the nature of the accounts and records which are to be kept, regulation 6(1) provides generally that these must be sufficient to show that the estate agent's basic duty in respect of clients' money has been discharged, and 'to show and explain readily at any time all dealings with the money'. This latter requirement involves identifying the account itself, the institution where it is located, and so on. From the point of view of the practising estate agent, however, the more important (and more useful) provision is regulation 6(3), which gives a detailed list of the information which must be made available.[2] In relation to clients' money received, this consists of the date of receipt; the amount; the name and address of the depositor; the purpose of the payment (eg pre-contract deposit, contract deposit, payment in respect of a connected contract);[3] the interest in land to which the payment relates (and the identity of the person wishing to dispose of that interest); and the person for whom, and the capacity in which, the money is from time to time held.[4] Further, it is provided that the majority of this information must also appear upon the counterfoils or duplicate copies of receipts issued to those who deposit clients' money, and that these counterfoils or copies must themselves be kept. Where payments out of the account are concerned, the information required to be shown is the date of the payment; the amount; the identity of the payee; the interest in land to which the payment relates; a reference to

---

1 Failure to keep such accounts is a criminal offence: s 14(8).
2 The information is to be 'indexed' be reference to the interest in land to which it relates: reg 6(3)(b).
3 If part of the sum is paid in only because it cannot be split, the purpose of that payment must be shown.
4 The regulation acknowledges that the estate agent may not know these matters, since both the person and the capacity may change while clients' money is in the agent's hands.

the corresponding payment in; and the purpose of the payment.[1] Finally, where money is transferred from one client account to another, the records must show the reason for the transfer and enable the original payment in to be traced. However, this last requirement does not apply where sums are transferred between a current and a deposit account kept by the agent, in both of which clients' money is held generally.

Audit

The basic obligation imposed by regulation 8 of the Estate Agents (Accounts) Regulations 1981 is to draw up the accounts and records described above in respect of consecutive 'accounting periods'[2] and to have them audited within six months of the end of each period by a 'qualified auditor'.[3] The auditor's report must then, under threat of criminal penalties, be produced on demand to a duly authorised officer of an enforcement authority.

The methods to be adopted by an auditor in this context are clearly defined by regulation 8. Having ascertained from the estate agent in question particulars of all bank accounts maintained by the agent or by employees in the course of estate agency work during the relevant period, the auditor is to examine the accounts and records kept by that person to see whether or not they comply with the requirements of regulation 6 described above. To this end the auditor may demand further information or explanations from the estate agent, but is not required to pursue enquiries beyond the accounts and records (as supplemented by these further explanations), nor to consider the keeping of the relevant accounts outside the period in question.

Having carried out this investigation, the auditor is required to produce a report to the estate agent, and this report may take one of four forms:

---

1   Where payment is made to the estate agent in respect of remuneration or expenses, the records must further identify any information required to be given to the client in respect of those charges under s 18.
2   A period not exceeding 12 months. No dates are specified, but an employee who is obliged to keep such accounts must adopt the same accounting period as the employer.
3   Defined by s 14(6) so as to mean, in nearly all cases, either a chartered accountant or a member of the Association of Certified and Corporate Accountants. In the case of limited companies, certain persons such as officers and servants of the company are disqualified from acting as auditors by the Companies Acts, and the Estate Agents Act applies the same disqualifications: s 14(7).

- that all the requirements of regulation 6 have been complied with
- that the requirements of regulation 6 have been substantially complied with. This would be appropriate where, in the auditor's opinion, the only breaches are trivial ones due to clerical errors, all of which were rectified on discovery and none of which have caused any loss
- that the requirements of regulation 6 have not been complied with. In this case the auditor must specify the relevant breaches and
- that the auditor is unable to form an opinion as to whether or not the requirements of regulation 6 have been complied with. In this case the auditor must specify the relevant matters and explain why it is impossible to form an opinion about them.

Interest on deposits
The position at common law

Under the general law of agency, any agent holding a client's money must account to the client for interest on that money.[1] This would certainly apply where an auctioneer or estate agent holds a contract deposit 'as agent' for the vendor. However, it seems that this principle does not apply to a deposit held by the auctioneer or estate agent 'as stakeholder', even where that deposit has remained in the auctioneer's possession for a considerable time. In *Lee* v *Munn*,[2] Burrough J said:

> An auctioneer can never be liable to interest, unless two circumstances concur. First, the contract must, on failure of the condition, be rescinded; secondly, a demand of the deposit must be made, and a refusal to return it must be given.

The Court of Common Pleas there made it clear that their decision rested upon the special circumstances of the case; in particular, the fact that the contract was kept open, and negotiations with the auctioneer pursued, for a period of four years. The general principle, however, was made the basis of an express decision of the Court of Exchequer in

---

1   *Brown* v *Inland Revenue Commissioners* [1965] AC 244.
2   (1817) 8 Taunt 45, 55. The decision was followed in *Curling* v *Shuttleworth* (1829) 6 Bing 121.

Gaby v *Driver*.[1] The vendor in that case having failed to make a good title, the auctioneer was sued for the return of the deposit with interest. It was held that no interest was payable, and Garrow B said:[2]

> I admit that when the contract has been finally broken off, he should return the deposit; but he has a right to notice that his character of stakeholder has ceased, before he can be charged personally with interest, on account of the detention of this money.

It should however be noted that, where the purchaser sues the vendor for breach of contract, the damages recovered will include loss of interest on the deposit while it is in the hands of the auctioneer.[3]

An auctioneer's immunity from liability to pay interest on deposit-money is even more far-reaching than has so far appeared. In *Harington* v *Hoggart*[4] it was held that there is no liability to account even for the interest which is actually earned by the money while it is in the auctioneer's possession as stakeholder. In that case the defendant auctioneer in 1813 sold the claimant's estate and received a deposit of £2,000, which he invested in a mixed fund with his own money. In 1816 the claimant, who was endeavouring to enforce the contract against an unwilling purchaser, instructed the auctioneer to invest the money for the vendor's benefit. When the purchase was finally completed in 1822, the vendor sued to recover the interest earned by the deposit-money. It was held that the auctioneer, as a stakeholder, was not liable to pay interest, and the express instructions given to him by the vendor did not alter the situation, for 'one party was not competent by himself to alter the contract on which the money was placed in the defendant's hands'.[5]

As to the reason for the auctioneer's immunity, it was suggested by Taunton J that, until one person became entitled to the deposit under the terms of the contract, the right to it was in abeyance, and that interest earned during this period could not be claimed. It is submitted, however, that the Court of King's Bench attached more importance to the potential liability of the stakeholder in respect of loss of part or all of the money. As Lord Tenterden CJ said:[6]

---

1   (1828) 2 Y & J 549.
2   At p 556.
3   *Farquhar* v *Farley* (1817) 7 Taunt 592.
4   (1830) 1 B & Ad 577.
5   At p 589, *per* Parke J.
6   At p 586.

If he think fit to employ it and make interest of it, by laying it out in the funds or otherwise, and any loss accrue, he must be answerable for that loss; and if he is to answer for that loss, it seems to me he has a right to any intermediate advantage which may arise.

To this may be added the view expressed by Harman J in *Smith* v *Hamilton*,[1] that the interest earned forms part of the stakeholder's remuneration.

Some inferential doubt was cast upon the principle laid down in *Harington* v *Hoggart* by the decision of the House of Lords in *Brown* v *Inland Revenue Commissioners*.[2] It was there held that, in the absence of express agreement, a solicitor was bound to account to clients for all interest earned on their money while it was in the solicitor's hands, no matter how small the individual amounts might be, or how great the difficulties of accounting. As Lord Upjohn said:[3]

A professional adviser, whether he be solicitor, factor, stockbroker or surveyor is of course in a fiduciary relationship to his client, and if and when he is entrusted with his client's money he can make no profit out of it.

In *Burt* v *Claude Cousins & Co Ltd*,[4] it was suggested by Sachs LJ that the *Harington* v *Hoggart* principle might require reconsideration in the light of *Brown* v *IRC*. However, some two years later that principle was not only approved but extended; in the case of *Potters* v *Loppert*,[5] Pennycuick V-C held that an auctioneer's freedom from liability to pay interest on a contract deposit was also enjoyed by an estate agent holding a pre-contract deposit. The legal basis of this decision was that there was no fiduciary relationship, in the *Brown* v *IRC* sense, between a stakeholder and a prospective purchaser; the stakeholder's only obligation was a purely contractual one to hand back precisely what was deposited. An ethical justification for this rule was found in the fact that, while holding a pre-contract deposit, an estate agent would have no prospect of earning commission by selling the property to anyone else; it would in effect be 'sterilised'.

---

1   [1951] Ch 174 at p 184.
2   [1965] AC 244.
3   At p 265.
4   [1971] 2 QB 426 at p 449.
5   [1973] Ch 399.

The Estate Agents Act 1979

The liability of a person engaged in estate agency work to account for interest on deposit money is now governed by section 15 of the Estate Agents Act 1979 and regulation 7 of the Estate Agents (Accounts) Regulations 1981. By providing that such account is to be made to the 'person who is for the time being entitled to the money', and specifically excluding any period during which the money is held 'as stakeholder on trust for the person who may become entitled to it or the occurrence of' a specified event, regulation 7(1) makes it clear that the same sum of money may attract a duty to pay interest to more than one person, in respect of different periods of time. As we have already seen,[1] the 'person entitled' to clients' money (and therefore entitled to interest on it) may be the actual or prospective purchaser (pre-contract deposit, or contract deposit once the sale has gone off); the vendor (contract deposit held 'as agent', or contract deposit after completion of the sale); or neither party (contract deposit held 'as stakeholder').

To avoid the creation of a legal obligation which might necessitate an expensive book-keeping operation in respect of a trifling sum, it is provided by regulation 7 that the duty to pay interest arises only where the clients' money in question exceeds £500 and the relevant interest is at least £10.[2] The relevant interest for this purpose is computed in different ways, according to where the clients' money is held.

- If the money is held in a separate deposit account (assuming of course that this is a valid client account), the estate agent must account for the interest which it actually earns.
- If the money is held in a client account which is not a separate deposit account, the agent must account for interest as if it had been so held at the same authorised institution.
- If (contrary to section 14 of the Act) the money is not held in a client account, the agent is penalised by having to pay interest at the highest rate available on the day when it should have been paid in.[3]

---

1   Pp 249–250.
2   This means, not the total interest earned by the sum in question, but the amount to which the particular person is entitled.
3   If the estate agent has one or more client accounts with authorised institution, payment must be made at the highest rate available at those institutions. If the agent has no such account, payment must be made at the highest rate available at *any* of the authorised institutions.

• Where an authorised institution is itself holding a deposit as an estate agent, interest must be paid at the appropriate rate for a separate deposit account at that institution.

Section 15 of the Estate Agents Act makes it clear that the obligations described above may be modified or excluded by an 'arrangement in writing, whenever made' between an estate agent and any other person who has or may have an interest in clients' money. Exactly what is meant by this is not clear. Presumably the mere giving of notice in writing (for example on the receipt which is issued to a depositor) falls short of being an 'arrangement between' the parties, but there is no positive requirement of signature by or on behalf of the person concerned. In the absence of such an arrangement, the statutory rules are practically exhaustive; section 15(3) provides that there is no other liability to account for interest on money held in a general client account.[1]

It should be noted that, unlike the other provisions of accounts regulations, those concerned with the duty to account for interest are not backed up by criminal sanctions. However, a breach may of course operate to trigger the powers of the Director-General of Fair Trading,[2] and it is also specifically provided by section 15(5) that any person entitled to interest on clients' money may claim it from the estate agent in the civil courts.

As a result of these statutory provisions, it seems that the common law rules will now be of relevance only in cases which fall outside the Estate Agents Act altogether, such as those where the transaction in question does not constitute 'estate agency work' as defined by section 1.[3]

---

1   This presumably does not preclude the possibility of the money being held in a *separate* client account, in such a case, it seems that any liability to pay interest is governed by common law in so far as it does not fall within the accounts regulations. If so, a liability *will* arise (even where the money is held 'as stakeholder'), since s 13 creates a trust where none existed previously.
2   See p 308.
3   See pp 270–285.

Payments other than deposits

Clients' money

As noted earlier in this chapter, the 'clients' money' rules imposed by the Estate Agents Act 1979 apply, not to clients' money as that term is commonly defined, but only to contract and pre-contract deposits. For the legal rules applicable to clients' money in the wider sense, one must turn to the general law of agency. This imposes a number of obligations upon every agent, which have been summarised as follows:[1]

> It is the duty of an agent —
> (1) where he holds money or property belonging to his principal, to keep it separate from his own and from that of other persons;
> (2) to preserve and be constantly ready with correct accounts of all his dealings and transactions in the course of his agency;
> (3) to produce to the principal, or to a proper person appointed by the principal, all books and documents in his hands relating to the principal's affairs.

Apart from the obligations imposed by law, those estate agents who are members of a professional body will invariably find that the rules of conduct which are a condition of membership contain obligations as to the correct handling of clients' money. A breach of such rules does not of itself attract legal sanctions, but renders the agent liable to disciplinary proceedings by the professional body, leading to whatever punishments (usually including expulsion from membership) are specified.

Of the professional bodies to which estate agents commonly belong, it is the *Conduct and Disciplinary Regulations* of the RICS[2] which contain the most comprehensive conduct rules in respect of clients' money.[3] For the purpose of these rules:

> 'Clients' money' means any money received or held by a Member or his practice over which he has exclusive control and which does not belong solely to him, or his practice or a connected person.

The RICS Commentary on this definition acknowledges that it is extremely wide and suggests, by way of illustration, that it includes:

---

1   *Bowstead & Reynolds on Agency* (17th ed), 6-088.
2   July 2001, Royal Institution of Chartered Surveyors, reg 27.7.
3   See also the *National Association of Estate Agents' Rules of Conduct* (July 1998).

- rents
- building society valuations fees (where the member is arranging the mortgage)
- service charges collected by a managing agent
- advance payments by clients of potential disbursements
- fees to be paid to other consultants
- fees paid by the client in advance
- moneys held as Law of Property Act receivers.

The RICS's detailed rules as to the treatment of clients' money are not dissimilar to those imposed by the Estate Agents (Accounts) Regulations 1981. They cover such matters as the designation of a client account, payments into and out of that account, the keeping of accurate records and a system for monitoring the member's handling of clients' money.

Before leaving the subject of professional conduct rules, one important point should be made. Where a deposit falls within the definition of 'clients' money' for the purposes of the Estate Agents Act, it must be dealt with in accordance with the provisions of that Act. This includes keeping it separate from all other money; not just the agent's own funds, but also any sums which might constitute 'clients' money' for the purposes of professional conduct regulations. It follows that an agent in this position must keep contract and pre-contract deposits in an account which is quite separate from any other client accounts.

Illegal premiums

Apart from deposits, there is one other area in which a client has been (controversially) held responsible for the receipt of money by an estate agent. This rests largely upon a single decision of the Court of Appeal, which is somewhat surprising since one suspects that the factual situation involved is not as uncommon as it might (or should) be. Nevertheless, although the case appears as an isolated landmark in the Law Reports and although certain views expressed by Denning LJ are open to criticism, the decision itself has been quoted with apparent approval on subsequent occasions.

The case in question, *Navarro* v *Moregrand Ltd*,[1] concerned the letting of a flat by an estate agent on behalf of the defendants. The claimant, who wanted a tenancy of the flat, was told by the agent that

---

1   [1951] 2 TLR 674.

he would have to make a payment of £225 in one pound notes, in addition to the rent. The claimant paid this sum to the estate agent, and then sued both landlords and agent to recover the money as an illegal premium under section 2(5) of the Landlord and Tenant (Rent Control) Act 1949. The County Court judge found that, although the agent had no authority from the landlords to demand such a sum, he had ostensible authority to conduct the full business of letting this particular property. Nevertheless he held that, when the agent demanded an illegal premium, the claimant received the clearest possible notice that the agent was exceeding his authority.

The claimant appealed to the Court of Appeal, where Somervell LJ took a very different view of the conclusions to be drawn from the evidence. Referring to the fact that the trial judge had treated the payment as a premium, and not as a bribe to the estate agent, he said:

> I think that the conclusion which anybody in the tenant's position would have drawn, dealing with a man placed in the position of the second defendant by the landlords, was that the landlords were demanding a premium for the granting of the lease.

The agent's illegal action was therefore within the scope of his ostensible authority, and the landlords were liable to the claimant for the sum paid.

Denning LJ reached the same conclusion as Somervell LJ, but by a completely different (and, it is submitted, quite erroneous) route. The estate agent, he said, had committed a 'statutory wrong' by demanding a premium for the granting of the lease. This wrong had been committed in the course of the agent's employment (and this would be so whether or not the agent had actual or ostensible authority) and the principal was therefore vicariously liable. This analysis of the law can be criticised on two main grounds. First, it represents an application of tort principles to what is in fact a case of quasi-contract, an action to recover a sum of money paid in contravention of a statute. Secondly, it reveals a confusion between the liability of an employer for the torts of an employee and that of a principal for the torts of an agent. General vicarious liability, with its allied concept of the 'course of employment', appears relevant only to the former relationship, where it is well established. No such broad principle covers the second situation where, if the principal is to be made liable, it must be established that the agent acted within the scope of authority.[1]

---

1   See P S Atiyah, *Vicarious Liability in the Law of Torts*, Chapter 9.

The precise status of *Navarro* v *Moregrand* today is a matter of some doubt. The judgment of Denning LJ has already been criticised; that of Somervell LJ, with its emphasis on the landlord's placing of the agent in a position to perpetrate the fraud, is strongly reminiscent of the argument in respect of pre-contract deposits which was rejected by the House of Lords in *Sorrell* v *Finch*.[1] Indeed, the latter decision renders it very difficult to claim that the agent in *Navarro* v *Moregrand* had implied authority to collect a premium, or even that such authority would be contained by implication within the ostensible authority to conduct the full business of letting.[2] In short, it may be best to regard the case as turning on its own individual facts which were sufficient to clothe the agent with the appearance of authority to do the very act complained of, namely, to collect an illegal premium.

---

1   [1977] AC 728: p 243.
2   This none the less appears to have been the view of the Court of Appeal in the similar case of *Saleh* v *Robinson* [1988] 2 EGLR 126, where *Navarro* v *Moregrand* was not mentioned.

The Estate Agents Act 1979

Notwithstanding its title, the 1979 Act falls a long way short of providing comprehensive statutory control of the estate agency profession. What is offered by the Act, and the package of Orders and Regulations made under the Act in 1991, is a number of limited controls upon the practice of estate agency; these controls fall a long way short of anything resembling an overall code.

The legislative background

The Estate Agents Act 1979 represents a return on more than 90 years of Parliamentary frustration for, between 1888 and 1978, there were at least 14 attempts by Private Members to secure statutory recognition for and control of the profession. There is little to be gained, in the context of our present enquiry, from an exhaustive analysis of these earlier failures,[1] but it is perhaps worth mentioning three threads which run with remarkable consistency through the story: the Bills presented all envisaged a positive registration scheme (with registration normally restricted to those holding a particular qualification); each scheme in turn, supported by one or more the major professional bodies of the day, was resolutely opposed by 'unattached' agents; and the professional bodies themselves were unable to achieve any real consensus over the most suitable form of control to adopt.

---

1   For the story up to 1967, see F A R Bennion: (1968) 206 EG 427, 543, 661, 779, 887.

A further feature of this period was that successive Governments were constant in their refusal to introduce legislation on the subject of estate agents, or to give any support (other than assistance at the drafting stage) to Private Members' Bills dealing with the matter. The lack of official interest thus demonstrated continued until 1975, when the Department of Prices and Consumer Protection published a Green Paper entitled: 'The Regulation of Estate Agency: a Consultative Document'. This found no evidence of frequent malpractice among estate agents, but referred to 'a continuing groundswell of complaint' about certain matters, notably some well-publicised cases involving the loss of deposits. The major suggestion contained in this document was that there should be established a scheme for licensing of any person seeking to carry on the business of estate agency, and that this should be administered, not by the profession itself, but by the Director-General of Fair Trading, on whose licensing functions under the Consumer Credit Act 1974 the new proposals were based.

The consultations which were stimulated by the Green Paper convinced the Labour Government that, while the time was ripe for statutory control of estate agents, the desired degree of consumer protection could be achieved without what the Minister termed 'a bureaucracy of control involving compulsory registration or licensing'. Oddly, detailed proposals then appeared, not in a Government Bill, but in the guise of a Private Member's Bill introduced by Mr Bryan Davies MP. This achieved a fair measure of support from both the main parties but was ultimately 'talked out' for political reasons. However, it served to pave the way for a Government Bill in almost identical terms which passed through Parliament in the following session and received the Royal Assent on 4 April 1979. Even then the delays were not over; the Estate Agents Act 1979 remained wholly inoperative until 3 May 1982, when most of the Act,[1] together with some supporting regulations, was brought into effect.

The scheme of control laid down by the 1979 Act was based upon what may be called 'negative licensing'. There was in general no requirement for any would-be estate agent to demonstrate any particular qualification or competence before setting up in practice;[2] the

---

1   Yet to be implemented are ss 16 and 17 (compulsory insurance cover for clients' money); s 19 (restriction on the size of pre-contract deposits); and s 22 (minimum standards of competence).

2   There is power under s 22 to impose such a requirement, but that section is not yet in force.

powers of the Office of Fair Trading (extending to the issue of an order banning an agent from practice altogether) would come into play only when the agent was shown to be unfit to continue, and when some other condition, such as a conviction for a relevant criminal offence, was also satisfied. True, the Act created a number of new duties with which estate agents must comply in their daily business but these, for the most part, operated only to 'trigger' the Office's powers.

In the first six years of its operation, the Estate Agents Act kept a fairly low profile. However, fundamental changes in the estate agency world, due both to the buoyant property market and to the incursions of major financial institutions, created a range of new pressures and new reported abuses. Official interest in estate agency was duly reawakened, and resulted in the publication of two reports within a period of six months. First came a review of the Estate Agents Act which was published by the Office of Fair Trading in December 1988. Then, in June 1989, the Department of Trade and Industry published its 'Review of Estate Agency'.

It must be said that neither the Office of Fair Trading nor the Department of Trade and Industry was unduly critical of the estate agency profession. The OFT report stated that the OFT had 'no evidence to suggest that the majority of estate agents fail to treat consumers fairly and efficiently'; the DTI document acknowledged that much of the public concern related to 'a perception of shortcomings in the property transfer system itself rather than estate agents'. Perhaps because of this, neither review was in favour of implementing any of the remaining provisions of the 1979 Act; the way forward, it was felt, lay in persuading estate agents themselves to produce a non-statutory Code of Practice. However, it was suggested that certain aspects of estate agency should be curbed by being designated 'undesirable practices' under the Act, and, moreover, that the Trade Descriptions Act[1] should be extended to cover land and buildings.

Responsibility for taking these proposals forward was given to the Office of Fair Trading. Action followed swiftly: a consultation document was produced in September 1989, and this led in March 1990 to a Final Report. This acknowledged defeat on a voluntary Code of Practice, confirmed an intention to extend the Trade Descriptions Act and contained draft proposals for the designation of a number of

---

1   Which, broadly speaking, governs false or misleading statements about goods and services.

'undesirable practices' (including, as an interim measure, one of property misdescription).

Within a mere 12 months, the necessary Orders and Regulations to be made under the Estate Agents Act had all been drafted, consulted over and laid before Parliament. These came into force on 29 July 1991. As for property misdescription, the Government's intentions in this area had been overtaken by a Private Member's Bill sponsored by the Consumers' Association and this, with Government support, duly received the Royal Assent on 27 June 1991. The statutory instrument upon which the Property Misdescriptions Act 1991 depended did not, however, appear until almost 18 months later, and the Act finally came into operation on 4 April 1993.[1]

The following description of the current statutory controls falls into three sections: the scope of the 1979 Act; the specific duties which the Act and supporting regulations impose upon practitioners; and the machinery by which these rules are enforced.

Scope of the Act

As mentioned at the start of this chapter, the Estate Agents Act 1979, even with its 1991 extensions, does not seek to provide a comprehensive system of statutory control for the estate agency profession as a whole. What the Act covers may be described as 'estate agency work in relation to freehold and leasehold interests in land, subject to a number of exclusions', a definition which must now be examined in some detail. It will be seen that many of the services commonly provided by estate agents (such as valuations or mortgage-broking) are nevertheless outside the scope of the 1979 Act, either because they do not satisfy the rather limited definition of 'estate agency work' or because they have been specifically excluded. As to precisely why these matters have been excluded, the detailed reasons vary, but the common thread is that in each case the element of 'consumer protection' involved was felt insufficient to justify the public expenditure required to extend the coverage of the legislation.

---

1   The Property Misdescriptions Act is considered in detail in Chapter 5.

Estate agency work

An estate agent is easy enough to recognise and fairly easy to describe; a satisfactory definition, however, is much harder to achieve, as the instigators of previous attempts at legislative control discovered. Rather than try yet again to attain this difficult objective, the Estate Agents Act adopts an alternative approach, namely, that of defining the estate agency function. As a result, this definition requires extremely careful study, since any person who carries out the functions described will be caught by the Act, whether or not he or she would otherwise be regarded as an 'estate agent'.

The crucial definition is contained in section 1(1), which provides:

> This Act applies, subject to subsections (2) to (4) below to things done by any person in the course of a business (including a business in which he is employed) pursuant to instructions received from another person (in this section referred to as 'the client') who wishes to dispose of or acquire an interest in land —

> (a) for the purpose of, or with a view to, effecting the introduction to the client of a third person who wishes to acquire or, as the case may be, dispose of such an interest; and

> (b) after such an introduction has been effected in the course of that business, for the purpose of securing the disposal or, as the case may be, the acquisition of that interest;

> and in this Act the expression 'estate agency work' refers to things done as mentioned above to which this Act applies.

The basic definition

It will be clearly seen from the above that the basic definition of 'estate agency work' is an extremely wide one. 'Things done' is not, perhaps, the most elegant of expressions, but it serves to indicate that the Act is capable of applying to any form of activity, provided that its object is either to effect an introduction between potential contracting parties or to bring such an introduction, once made, to a successful conclusion. Furthermore, the specific reference to both acquisitions and disposals of interests in land makes it clear that the Act covers estate agents, not only where they act for vendors or landlords who are seeking to dispose of property, but also where they represent purchasers or tenants who are seeking to acquire it.[1] However, it may be noted that certain of the statutory duties are worded in such a way that they can only apply to agents acting for vendors or landlords.

As to the first of the statutory alternatives (the effecting of introductions), few problems of definition seem likely to arise, especially given that persons who do nothing more than publish advertisements are specifically excluded from the operation of the Act by section 1(4). Paragraph (b), however, leaves a curious loophole in relation to post-introduction work, by providing that such work is only covered when it takes place in the course of the same business as the introduction which it follows. If it were not for this restriction, it might perhaps be argued that building societies agreeing to lend money on mortgage, or Land Registry clerks dealing with searches, were doing things for the purpose of bringing transactions to fruition and were therefore within the scope of section 1(1). Common sense suggests that such persons should not come within the Act; however, in seeking to prevent them from doing so, the wording also serves to exclude the work of an estate agent who is instructed to negotiate a sale or lease on behalf of one of the parties, at a time when the introduction has already been made by another agent or by the parties themselves. Such work is not uncommon; indeed, the meaning of 'negotiation' in this context has twice received the attention of the Court of Appeal.[2] Nevertheless, and despite the fact that such an agent, if called in to advise a vendor, might well receive a deposit from the purchaser (control of which is one of the primary objectives of the Act), it is quite clear that this situation is not covered by the statutory definition of 'estate agency work'.

Business

Many of the introductions which lead to a sale of property are made by a mutual acquaintance of the parties on a purely social basis. It would not of course be desirable to subject these to the provisions of the Act and, one might have thought, a simple way to avoid this would be to confine the operation of the statute to cases where the

---

1   An estate agent who acts for a person seeking the tenancy of a dwelling house is subject to the Accommodation Agencies Act 1953: see p 331.
2   *Rolfe (FP) & Co* v *George* (1969) 210 EG 455; *Hoddell* v *Smith* (1975) 240 EG 295. See p 124.

person concerned is paid for the introduction. This, however, has not been done; the Act applies to all work, whether paid or not, provided only that it is done both 'in the course of a business' and 'pursuant to instructions received'.

The first point to be noted in relation to this chosen form of words is that the 'business' in question need not be an estate agency business; indeed, it need have little or no ostensible connection with the property world at all. For example, it has been said that bank managers not infrequently mention to one client who is seeking to sell a house the name of another client who may be interested in buying it, and there can be no doubt that any such introduction would fall within the course of the bank manager's business. The same would of course be true of members of other professions, such as accountants, who might well happen to introduce two of their clients and thereby indirectly bring about a sale of property.

As to the meaning of 'business' itself, the Act provides no definition. If previous decisions of the courts upon other legislation may be relied upon as a guide, it seems that a 'business' connotes both a degree of continuity (ie more than one isolated transaction) and the general idea of gain or reward. However, it is not essential that a person intends to make a gain out of each and every transaction, nor need the 'business' show an overall profit; thus, for example, any activities of a local authority, the Housing Corporation or a housing association which satisfy the rest of the definition of 'estate agency work' will come within the Act.

Instructions

It is not every introduction made by a person in the course of business that will constitute 'estate agency work' for the purposes of section 1, but only those which are effected 'pursuant to instructions received'. 'Instructions', like 'business', are not defined by the Act, but it seems clear that they may quite validly be given by word of mouth, whether or not they are subsequently confirmed in writing. Nonetheless, there must always be some element of instruction, however informal; if this is lacking, then the person concerned is not engaged in 'estate agency work'. As a result, the Act does not apply to sales of land effected by such people as the personal representative of a deceased landowner, a trustee in bankruptcy or the liquidator of a company; all these 'officials' act by virtue of their position, rather than pursuant to instructions received.

It has been suggested that the concept of 'instructions' will serve to exclude from the Act those other professional persons, such as bank managers and accountants, mentioned above, who may in the course of business introduce one of their clients wishing to sell a house to another client wishing to buy one. No doubt this is true of some cases, where the intermediary acts on his or her own initiative, but equally there must be many occasions where a bank manager is specifically asked by a prospective vendor to keep an ear to the ground, and here the requirement of 'instructions' would seem to be satisfied.

Employees

Much of an estate agent's work is carried out by employees, who may (depending on their type and status) be remunerated on a salary basis or by the payment of commission on business done. Section 1(1) makes it quite clear that such an employee is engaged in 'estate agency work' and is therefore subject to all the provisions of the Act and its supporting regulations. Indeed, it is not necessary for a person to be acting pursuant to instructions which that person has received from a client; the instructions may have been received by the employer or by a fellow-employee.

The effect of this is to bring every member of an estate agent's staff[1] within the Act. As a result, each employee must comply with all the various statutory duties;[2] any failure to do so may lead to investigation of the individual employee by the Office of Fair Trading, with all the consequences that this may involve. In practical terms, such coverage is essential in order to provide for the case of a firm with branch offices, some of which may be under the sole control of a person who is an employee rather than a principal in the business.

Not only is the employee of an estate agency firm subject to the provisions of the Act; responsibility for what that employee does in the course of business may also be placed upon the firm itself. This may involve criminal liability[3] or, under other legislation,[4] a finding that

---

1 At least all those who are employed under 'contracts of employment': see s 1(5)(c).
2 Except the duty to open a 'client account'; s 14(1) permits an employee to pay clients' money into a client account maintained by the employer.
3 There may be vicarious liability in respect of crimes which do not require the prosecution to prove guilty intent.
4 Sex Discrimination Act 1975, s 41; Race Relations Act 1976, s 32.

the firm is guilty of discrimination. Further, in connection with the events which empower the Office of Fair Trading to make an order prohibiting an estate agent from practising, section 3(3)(a) imposes vicarious liability upon the employer in respect of those duties under the 1979 Act which do not lead to criminal sanctions and also any 'undesirable practices' as defined by statutory instrument.[1]

Excluded matters

Section 1(1) defines 'estate agency work'; the remainder of section 1 specifies no fewer than nine situations in which work apparently satisfying this definition is nevertheless not covered by the Act. Although these exclusions are not classified by the Act itself, they may be seen as falling into two groups. In the first place, there are a number of things which may well be done by an estate agent, either in connection with the disposal of a particular property or quite independently of it, which have been thought for one reason or another not to warrant statutory control. Second, certain work carried out by persons who are not themselves 'estate agents' is also excluded.

Things done by estate agents

Section 1(2)(e) excludes from the operation of the Act things which are done 'in connection with applications and other matters' which arise under current planning legislation. The reason for this exclusion, which appears to cover such advice even where it is given in the course of 'estate agency work', is not immediately apparent. Presumably it was felt by those responsible for drafting the Act that the need for consumer protection, which was its motivating force, was much less evident in this connection than in relation to those functions more directly connected with the actual disposal of property.

Things done 'in the course of carrying out any survey or valuation pursuant to a contract which is distinct from that under which other things falling within subsection (1) above are done' are excluded by section 1(2)(d). This again gives a sharper focus to 'estate agency work' by making it clear that it is only when such work is already being done that an incidental survey or valuation is included within the definition.

---

1   S 3(1)(c) and (d): see p 314.

This, it is suggested, would cover the case where a valuation forms par of the service which is offered by an estate agent to a prospective vendor. It would not, however, apply to a mortgage valuation carried out on behalf of a building society, even if the valuer's firm had been responsible for introducing the property to the borrower concerned. What, then, of the case where an agent retained by a prospective purchaser, having found a suitable property, then carries out a structural survey of that property on the client's behalf? This is more doubtful, but it would probably be held to fall outside the scope of the Act, as such a survey would normally be done under a separate contract.

'Estate agency work' does not include things done 'in the course of credit brokerage, within the meaning of the Consumer Credit Act 1974'.[2] The reference is to section 145(2) of that Act, which deals with 'the effecting of introductions — in the case of an individual desiring to obtain credit to finance the acquisition or provision of a dwelling occupied or to be occupied by himself or his relative, to any person carrying on a business in the course of which he provides credit secured on land'. This clearly applies to the common case of an estate agent who arranges mortgages, and such a person will therefore need to be licensed under Part III of the 1974 Act as carrying on an 'ancillary credit business'. In view of the strict control which this involves, and the supervisory functions exercised by the Office of Fair Trading, to have brought credit brokerage within the Estate Agents Act as well was seen as an unnecessary duplication. It should be noted, however, that mortgage-broking in respect of non-residential premises is beyond the scope of the Consumer Credit Act; as a result, an estate agent carrying out this particular function remains within the Estate Agents Act, provided of course that the agent's firm is responsible for introducing the parties.

Similar reasoning underlies section 1(2)(c), which excludes things done 'in the course of insurance brokerage by a person who is for the time being registered under section 2, or enrolled under section 4, of the Insurance Brokers (Registration) Act 1977'.[3] It should be noted, however, that this provision is of much narrower scope than that relating to credit brokerage. The 1977 Act does not compel all those

---

1   If the firm had not effected the introduction, the Act would in any case not apply, since the valuation would not then follow an introduction made in the course of the agent's business.
2   S 1(2)(b).
3   Registration applies to individuals, enrolment to companies.

who practise insurance broking to register or enrol; it applies only to those who wish to use the title of 'insurance broker'.[1] Hence, if an estate agent arranges insurance, without using a restricted title or falsely claiming to be registered, these activities will be governed, not by the Insurance Brokers (Registration) Act, but by the Estate Agents Act.

Things done by other persons

Five groups of persons are specifically excluded from the statutory controls. In the first place, things which are done 'in the course of his profession by a practising solicitor or a person employed by him' are excluded from the operation of the Act by section 1(2)(a). Since the normal conveyancing work of a solicitor, although undoubtedly of assistance in bringing about a completed transaction, does not follow an introduction made in the course of the solicitor's business, this would not in any case be covered by the Act. Section 1(2)(a), therefore, is concerned to exclude cases where a solicitor is in effect acting as an estate agent, for example by operating a 'property shop'. This apparently privileged position was heavily criticised by estate agents' professional bodies, especially in its application to Scotland, where solicitors have traditionally played a much more active role in the actual selling of property and have therefore been in direct competition with estate agents. The Government's response to such criticism has been to argue that existing controls on solicitors[2] are at least as rigorous as those imposed upon estate agents and that to subject solicitors to both régimes would, from the point of view of consumer protection, be superfluous. Whether or not the Government view is correct depends upon one's assessment of the relative burdens imposed by two different sets of regulations. However, it is interesting to note that one of the main grounds of estate agents' complaints (that solicitors were not obliged to give their clients advance warning of charges for estate agency services) has since been remedied by a Guidance Note from the Council of the Law Society, which requires a written agreement signed by the client at the time when instructions to negotiate a sale are given and accepted.

---

1   For the precise nature of the restrictions which are imposed upon unregistered persons, see the Insurance Brokers (Registration) Act 1977, s 22.

2   S 1(5)(a) defines solicitors in such a way that those who are not subject to their own professional code are caught by the Estate Agents Act.

Second, in order to avoid an unnecessary extension of statutory control to newspapers, computerised house-finding agencies and the like, whose activities in the estate agency field consists solely of the provision of information, it is laid down by section 1(4) that the Act does not apply to 'the publication or the dissemination of information by a person who does no other acts which fall within subsection (1)'. However, the exemption applies only where the organisation goes no further than merely providing information; any further measures designed to effect introductions for clients are governed by the Act.

Thirdly, as a result of anxiety expressed by accountants through their professional bodies, it was suggested that the Estate Agents Act should specifically exclude the activities of liquidators, trustees in bankruptcy, executors and so on who might, in that capacity, sell property belonging to another person. The official view, however, was that, with one exception, such a person would be acting by virtue of an office rather than 'pursuant to instructions' and would not therefore in any case be caught by the Act. The exceptional case (which, it was felt, might come within the definition of estate agency work but ought not to) was dealt with by section 1(3)(b), which excludes things done by any person 'in relation to any interest in any property if the property is subject to a mortgage and he is the receiver of the income of it'.[1]

Fourthly, section 1(3)(a) excludes from the operation of the Act things done by any person 'pursuant to instructions received by him in the course of his employment in relation to an interest in land if his employer is the person who, on his own behalf, wishes to dispose of or acquire that interest'. The purpose of this provision may have been simply to avoid the creation of a client-agent relationship between an estate agent and an individual employee[2] where that estate agent seeks to buy or sell a property through the firm, but the effect of the statutory wording appears to be wider than this in at least two respects. In the first place, it seems that, while an estate agent who actually has a personal interest in property will be in breach of section 21 of the Act[3] if this is not disclosed in any negotiations, the employee who actually conducts those negotiations with a third party will owe no such duty of disclosure. This could of course lead to the removal of the Act's protection in the very circumstances in which it is most

---

1   'Mortgage' is defined by s 1(5)(b).

2   It is clear that an employee is, potentially at least, subject to all the statutory controls: see p 274.

3   And of the Estate Agents (Undesirable Practices) (No 2) Order 1991: see p 297.

needed. Second, it is at least arguable that the exemption may also apply where the 'instructions' mentioned in section 1(3)(a) come, not from the employer, but from an independent client, and where the employee then carries out those instructions by selling to his or her own employer. This can hardly have been intended, and it is to be hoped that a court would resist this conclusion by interpreting the phrase 'the person who, on his own behalf, wishes to dispose of or acquire that interest' as referring back to the person described as a 'client' in section 1(1).

It is not only the employees of estate agents who may be removed from the operation of the Act by section 1(3)(a), but anyone carrying out what would otherwise constitute 'estate agency work' on behalf of his or her own employer.[1] This means, for example, that office workers may safely assist the employer in arranging a 'private' sale of the employer's own house without finding that they have inadvertently become estate agents. More importantly, where a firm of builders or developers sells its houses direct to the public, the activities of its employees in negotiating sales, collecting deposits, etc are outside the scope of the Estate Agents Act.[2]

Fifth, and less controversially, section 1(3)(c) excludes things done by any person 'in relation to a present, prospective or former employee of his or of any person by whom he also is employed if the things are done by reason of the employment (whether past, present or future)'. As a result of this provision, a firm which assists its employees in buying or selling their own houses will not be treated as carrying on 'estate agency work', and nor will the specific employees whose job it is to provide that assistance. The exemption also covers cases where the 'employee' assisted has not yet started work or has ceased to work for this employer (eg where he or she is moving into the area to take up the employment or leaving the area at the end of it).

Interests in land

The Estate Agents Act was originally intended to apply only to estate agency work which related to residential property, or at least to property with a substantial residential element. To critics who

---

1   Provided that the relationship between them is a true 'contract of employment': see s 1(5)(c).

2   Though they will be subject to the Property Misdescriptions Act: see p 221.

regarded the buyer of a small business as being as much in need of 'consumer protection' as the buyer of a house, the Government made two replies: first, that 'small' business deals were difficult to define and there was no evidence to suggest that large ones caused any problems; and, second, that the consultations which had been carried out prior to the introduction of the Bill had dealt only with residential property. Unconvinced by these explanations, however, the critics redoubled their efforts to extend the Bill's coverage beyond the residential sector, and these efforts ultimately proved successful during the committee stage in the House of Lords. Political events then denied the Government the opportunity to restore the earlier restriction, so that the legislation in its final form applies to all forms of property, whether residential, commercial, industrial or agricultural or any combination of these.[1] There is, moreover, no mention of price, so that the Act will apply even to a wholly gratuitous transfer of land, provided that an estate agent is involved in arranging it.

Section 2 sets out the particular interests in land whose acquisition or disposal attracts the operation of the Act. The legislation is not concerned with minor interests in land, such as rights of way or mortgages;[2] the basic concept is that of ownership. First and foremost, therefore, the Act applies to the transfer of 'a legal estate in fee simple absolute in possession', that is, to the simple case where a freehold is marketed. About this little need be said, except to note that the restriction to legal estates will exclude sub-sales of property, at least those in which a person is selling the right to acquire the property directly from the original vendor, rather than undertaking to acquire the property and sell it on.[3]

If freehold transactions are relatively straightforward, the same unfortunately cannot be said of those involving the creation or transfer of leases. The first point to be made is that the absence of the word 'legal' in section 2(1)(b) (by contrast with section 2(1)(a)) suggests that equitable as well as legal leases are included. In fact, even if this interpretation were not accepted, those equitable leases which are

---

1 Membership of an approved redress scheme is, however, compulsory only for estate agents dealing with residential property: see p 338.

2 S 2(3)(a) makes it clear that this is so, even where the mortgagee's security consists of a leasehold interest in the property under the Law of Property Act 1925, ss 85 and 86.

3 In the former case what the sub-purchaser receives is an equitable right to call for a conveyance of the property.

deemed to arise where the parties fail through lack of the proper formalities (ie use of a deed) to create a legal lease[1] would in any event be brought within the Estate Agents Act by section 2(2), which provides that 'the expression 'lease' includes the rights and obligations arising under an agreement to grant a lease'.

The major problems in respect of leases arise from the Government's belief that the Estate Agents Act should not apply to agents involved in arranging rentals or property management, since it was felt that other statutes governing such activities already offered sufficient in the way of consumer protection. Attempting to define precisely what was to be excluded from the Act was, however, no simple task, and the formulation finally adopted sought to make the crucial question whether or not the lease was one for which a premium could lawfully be charged on either creation or assignment.[2] As we shall see, this formulation may have the unintended effect of excluding leases of any property at a full market rent.[3] However, whether or not this is so, the very idea of defining the scope of the Estate Agents Act by reference to the lawfulness or otherwise of premiums today bears an irrelevant, indeed rather arbitrary, air. This is because the prohibition on requiring or receiving a premium on the grant or assignment of a protected tenancy under the Rent Act 1977 does not extend to an assured or assured shorthold tenancy under the Housing Act 1988. In consequence, the vast majority[4] of residential tenancies created after 15 January 1989 will not be excluded from the operation of the Estate Agents Act by this provision. However, it remains generally unlawful[5] to require or receive a premium on the assignment of a protected tenancy (which means most residential tenancies created before that date) and thus an estate agent involved in the assignment of such a tenancy is not subject to the Estate Agents Act.

---

1   Under the principle of *Walsh* v *Lonsdale* (1882) 21 ChD 9. The vast majority of equitable leases are of this kind.

2   This effectively excluded from the Estate Agents Act the vast majority of transactions involving property within the Rent Act 1977, though that Act does permit the taking of a premium in certain cases (mainly involving assignments).

3   See p 285.

4   In exceptional cases it may still be possible to create a protected tenancy: see for example the Housing Act 1988, s 34.

5   There are some exceptions, most of which are listed in the Rent Act 1977, s 120.

Doubtful areas

The discussion in this chapter of sections 1 and 2 of the Estate Agents Act has highlighted a number of potential uncertainties in the definition of 'estate agency work', and thus in the scope of the legislation. We may now examine several areas of estate agency activity where, for various reasons, the applicability of the Act is a matter of some debate.

Sales by auction or tender

As we have seen, section 1(1) defines estate agency work by reference to the 'introduction' of two parties (the client and another) who wish respectively to acquire and to dispose of an interest in land. It might be suggested that this form of words serves to exclude both auctions and sales by tender from the operation of the Act, on the ground that what is brought about by the agent in such cases is not an introduction of the prospective vendor and purchaser, but rather an immediate contract between them. However, it seems clear that the Act was intended to cover sales of land by whatever method, and it is certainly desirable that it should do so, if only because of the difficulties which might arise where, for example, an agent is instructed to sell land 'by auction or otherwise'. In any event, it may be suggested that the term 'introduction', in the sense of bringing parties together, is wide enough in meaning to include these procedures.

The view that auctions at least are covered by the legislation is certainly shared by those responsible for drafting the Estate Agents (Undesirable Practices) (No 2) Order 1991, which provides a specific exemption for auctioneers from one of the duties which it imposes. None the less, it must be acknowledged that some of the specific estate agency rules do not sit too easily in the auction context. In particular, auctioneers may find severe difficulty in complying with the requirement that every offer received for a property is to be communicated to the vendor 'promptly and in writing'![1]

Overseas property

There is a basic presumption, rebuttable only by very clear words, that the effect of an Act of Parliament does not extend beyond the United

---

1   See p 296.

Kingdom. In consequence, the legislation now under discussion can only apply to 'estate agency work' which is carried out in England, Wales, Scotland or Northern Ireland. However, a separate question, which the presumption does nothing to answer, is whether the Estate Agents Act can apply to UK agents who market overseas property. This is a common enough activity, and it offers as much scope for abuse as does domestic property; none the less, it is doubtful to say the least whether it falls within the Act.

The view that sales of overseas property lie outside this legislation[1] is based largely upon section 2 of the 1979 Act. This section, in describing the interests in land which are covered, adopts the language of English (and Scottish) land law, using phrases such as 'legal estate in fee simple absolute in possession' and 'Register of Sasines'. Such terms may well be inappropriate when referring to immoveable property in other countries whose land law is fundamentally different from that of the United Kingdom, which suggests that the Act cannot apply to such property.[2]

Although the view outlined above is probably the correct one, it should be pointed out that it is based upon inference, rather than upon any express provision. What is more, some support for the opposing view, that the Act does in fact extend to UK transactions involving overseas property, may be extracted from section 12. This defines 'clients' money' as deposits relating to 'an interest in land in the United Kingdom'. Why, one might ask, is it necessary to place this qualification on the definition in section 12, if the entire Act only applies to UK property? If nothing else, the existence of such doubts leads to the practical conclusion that estate agents handling overseas sales should adhere to the same principles as they adopt in relation to their domestic work.

Sub-agency

The question of how far, if at all, the estate agency legislation applies to sub-agents is a surprisingly complex and difficult one to answer. It is clearly desirable that sub-agents should be subject to the statutory

---

1   A view which appears to be held by both the Department of Trade and Industry and the Office of Fair Trading.

2   The wording of section 1(6) of the Property Misdescriptions Act 1991 seems even more clearly to rest on this assumption.

controls, especially those which seek to protect members of the public rather than clients, since the various abuses at which these are aimed are just as capable of arising at the sub-agency level. However, it is clear that the legislation has not been drafted with sub-agency specifically in mind, and there is consequently some difficulty in bringing it within the statutory wording.

The basic problem lies in the 1979 Act's definition of 'estate agency work', which requires the agent's instructions to have been 'received from another person (in this section referred to as 'the client') who wishes to dispose of or acquire an interest in land'. In considering whether these words apply to sub-agents, the most natural interpretation is surely that a sub-agent's 'instructions' are received from the main agent, and that the main agent is not a person who wishes to dispose of or acquire an interest in land. If this is correct, then what the sub-agent does is not 'estate agency work' as defined by the Act.

In order to bring sub-agents within the Act (and, for that matter, within the Property Misdescriptions Act), it might be argued that the main agent is a person who wishes to dispose of or acquire land, albeit only as an agent for someone else. This argument, if accepted, would lead to the conclusion that the sub-agent's 'client' is the main agent, and that all the statutory controls therefore apply 'one step down'. The conclusion is attractive, but the argument itself does not seem very convincing.

A third possible interpretation of the legal position would be to treat the sub-agent as working pursuant to instructions which emanate from the real client (ie the vendor or purchaser) and which are merely transmitted to the sub-agent through the main agent. This avoids the need to strain the meaning of the word 'client' in section 1; unfortunately, it creates a situation in which the sub-agent and the 'client' do not have a direct contractual relationship with each other. This in turn leads to the conclusion that, while most of the statutory obligations would apply to a sub-agent, the duty to give the client written notice of charges could not, since that duty requires a direct contract between the 'agent' and the 'client'.[1]

---

1   S 18(1): see p 292.

Leases at a rack rent

As has already been pointed out, the original intention was that the Estate Agents Act should apply in principle to transactions involving leasehold interests, while excluding residential lettings and property management. However, the chosen method of expressing this distinction raises the uncomfortable possibility that the baby may have been thrown out with the bath water. The problem lies in section 2(1)(b), which applies the Act to 'a lease which ... has a capital value which may be lawfully realised on the open market'. Had this description been: 'a lease whose capital value if any may be lawfully realised', it would have indicated clearly that what mattered was whether any premium value could legitimately be extracted, rather than whether there was any such value to extract. As it stands, however, the wording of this provision suggests that, where a lease has no capital value (because the property is let at full rental value), no transaction involving that lease can fall within the Estate Agents Act.[1]

If this is indeed the true position, it seems to introduce an unnecessary and unproductive complication into what should be a simple matter of deciding whether or not the Act applies.[2] The crucial question, be it noted, is not whether a premium has been paid as part of a transaction, but whether the lease actually has a capital value. And this question is of course one which may receive different answers at different times, as fluctuations in the property market determine whether or not the current rent is a rack rent.[3]

It can hardly have been intended that an estate agent should have to carry out a complex valuation exercise in order to discover whether or not the Estate Agent Act applies. However, unless a somewhat extended interpretation can be given to the term 'capital value',[4] that appears to be the inevitable consequence of section 2(1)(b).

---

1   Unless, perhaps, it can be argued that any lease has in principle a capital value, albeit one which, depending on the rent charged, may be 'nil' at any given time.

2   The problem would of course have been far less serious if the Act, as was intended, had been limited in scope to residential property.

3   In a falling market, as agents will be painfully aware, leases can rapidly acquire a negative value.

4   See note 1.

Duties of estate agents

From the point of view of the practising estate agent, the most important aspect of the Estate Agents Act 1979, and the Orders and Regulations made under it, is the imposition of detailed rules governing the way in which day-to-day estate agency business is carried on. The legislation creates a number of specific duties which are owed, not only to clients, but also to third parties with whom an estate agent deals, and breaches of these duties can lead in appropriate cases to penalties ranging from loss of commission through a criminal prosecution to an order from the Office of Fair Trading which bans the agent altogether from continuing to practise.

In considering the specific statutory duties of estate agents, we shall not follow the order in which these duties appear in the legislation, but rather the approximate order in which they may be expected to affect a practitioner. Thus, in a straightforward case, an estate agent on obtaining instructions from a client must immediately furnish that client with certain information, concerned largely with the agent's charges. When the property is put on the market, the agent must keep the client informed about certain specified matters, and must also deal in a fair manner with any person with whom negotiations take place. If negotiations reach the point at which the agent is in receipt of deposit money (including a pre-contract deposit), this must be dealt with in accordance with strict statutory rules. We conclude with a reminder that, while the right to practise estate agency is in principle open to anyone, whether qualified or not, there are some special restrictions on that right.

Membership of an approved redress scheme

Section 23A of the Estate Agents Act[1] gives power to the Secretary of State to make regulations requiring some or all estate agents to be members of an approved redress (or 'Ombudsman') scheme. That power has been used to issue the Estate Agents (Redress Schemes) Order 2008,[2] under which membership of such a scheme is compulsory for any person engaging in estate agency work in relation

---

1   As inserted by s 53 and Schedule 6 of the Consumers, Estate Agents and Redress Act 2007.

2   SI 2008 No 1712, in force from 1 October 2008.

to residential property.[1] Failure to comply with this requirement is a 'trigger' for the OFT's power to ban an agent from practice; it is also something for which an enforcement officer may issue a 'penalty charge notice' (effectively a fine to be paid within 28 days), the level of which is initially set at £1000.[2]

It is for the Office of Fair Trading to decide whether or not to 'approve' a particular redress scheme, on the basis of requirements laid down in a new Schedule 3 to the 1979 Act. The OFT must be satisfied that a proposed scheme makes satisfactory provision about:

- the range of complaints that can be made under it
- the powers and duties of the ombudsman in relation to their investigation and determination
- the range of redress available (which must include at least include the provision of an apology or explanation and the payment of financial compensation) and
- the enforcement of the ombudsman's decisions.

At the time of writing, two such schemes had been approved, the already existing Ombudsman for Estate Agents scheme[3] and the one administered by Surveyors Ombudsman Services Ltd for agents who are RICS members.

Information to clients

In an effort to bring about the maximum possible transparency in dealings between estate agents and their clients, section 18 of the Estate Agents Act requires certain information to be provided by the estate agent as soon as an agency relationship is in view.[4] That requirement has been significantly extended by the Estate Agents (Provision of Information) Regulations 1991 in respect of both the information which is included and the time and manner at which it is

---

1 The meaning of 'residential property' for this purpose is set out in s 23C.
2 S 23B.
3 See p 337. It appears likely that the great majority of estate agents will choose to join this scheme.
4 In principle, the requirement applies as much to agents acting for prospective purchasers or tenants as to those acting for vendors or landlords. However, some of the 1991 provisions make sense only in relation to vendors' agents.

to be provided. It should be appreciated, however, that neither the Act nor the Regulations seek in any way to restrict the terms on which an estate agent contracts with the client; all that is required is that those terms are properly set out at the correct time.

Details of charges

The main thrust of section 18 is geared to information about all the charges, whether in the nature of remuneration or otherwise, for which the client may become liable. This information, according to section 18(2), comprises:

(a) particulars of the circumstances in which the client will become liable to pay remuneration to the agent for carrying out estate agency work;

(b) particulars of the amount of the agent's remuneration for carrying out estate agency work or, if that amount is no ascertainable at the time the information is given, particulars of the manner in which the remuneration will be calculated;

(c) particulars of any payments which do not form part of the agent's remuneration for carrying out estate agency work or a contract or pre-contract deposit but which, under the contract referred to in subsection (1) above, will or may in certain circumstances be payable by the client to the agent or any other person and particulars of the circumstances in which any such payments will become payable; and

(d) particulars of the amount of any payment falling within paragraph (c) above or, if that amount is not ascertainable at the time the information is given, an estimate of that amount together with particulars of the manner in which it will be calculated.

If all this information is duly given, a client will know at what point the agent's remuneration is to become payable (whether this is on the mere introduction of a prospective purchaser, the exchange of contracts, completion of the sale or some other specified event), and either the precise amount of that commission or the method by which it is to be assessed (eg professional scales or percentage of the purchase price). It has further been accepted by a Scottish court that, where an estate agent obtains a 'secret profit' from the agency work which is concealed from the client (consisting in that case of a discount which the agent was given on newspaper advertising) the agent is also guilty of a breach of section 18(2)(a), since this profit constitutes part of the agent's 'remuneration'.[1]

Compliance with section 18(2) also means that the client will be informed about any potential payments which do not amount to commission, such as advertising or other out-of-pocket expenses.[2] In relation to such payments, the agent must state either the exact amount which is to be charged or, where this is not possible at the outset, the method of its assessment together with an estimate. The Office of Fair Trading's view of this provision is that it requires the agent to provide as far as practicable an itemised list of expenses for which the client will be responsible, together with an estimate of the charges which will arise under each heading. It is emphatically not regarded as sufficient for the agent simply to agree a global maximum figure for expenses with the client, even if the agent then keeps within that figure.[3]

Special commission clauses

One of the main criticisms of estate agents, in the period leading up to the 1991 legislation, lay in their use of certain restrictive terms of business which might prove unduly onerous to inexperienced clients. Three such terms ('sole agency', 'sole selling rights' and 'ready, willing and able purchaser') came in for especially heavy criticism, and the Government's response was to require any estate agent using these terms to provide the client with an explanation of their meaning. Not only that; regulations 5 and 6 of the Estate Agents (Provision of Information) Regulations 1991 prescribe the exact words in which the explanation must be couched, subject only to the proviso that, if the circumstances render a statutory explanation misleading, it is the estate agent's responsibility to amend it so as to ensure accuracy.[4]

---

1   *Solicitors Estate Agency (Glasgow) Ltd* v *MacIver* 1993 SLT 23: see p 294.
2   A payment for abortive work (such as a 'withdrawal fee') which exceeds the expenses actually incurred will be regarded as 'remuneration' and will therefore fall within paragraphs (a) and (b).
3   Warning Orders issued on this ground against the firm of Burling Morrison and one of its partners were overturned on appeal to the Secretary of State. However, on 8 September 1988 the Office of Fair Trading issued a press notice stating that, having taken counsel's opinion, it maintained its original interpretation (which furthermore had been accepted by the RICS and the NAEA).
4   The wording of each of these definitions is considered in Chapter 3: pp 129–131.

It might be thought that the new rules could be evaded by the simple expedient of using variations on these terms, such as 'exclusive agency' or 'person willing to purchase'; however, this potential loophole has been foreseen and duly closed. The 1991 Regulations provide that, if an estate agent uses any terms which 'have a similar purport or effect' to those listed, the appropriate statutory explanation (again amended as far as necessary to avoid any misunderstanding) must be used.

The clear intention of regulations 5 and 6 is that the relevant statutory explanation should appear in whatever document is used by the agent to satisfy the general requirements of section 18. In any event, the time within which this information is to be given to the client is the same as for all information relating to the agent's terms of business. Furthermore, it is specifically provided that the statutory explanations must be given no less prominence than any other information in the document in which they are contained.

Services offered to purchasers

One area where the Estate Agents (Provision of Information) Regulations 1991 break entirely new ground is in connection with the provision, by estate agents acting on behalf of vendors, of 'tie-in' services to prospective purchasers. This practice, which is perhaps especially prevalent where the estate agent concerned forms part of a larger organisation (such as a financial institution) has given rise to a range of complaints, some of which have now been addressed in the Estate Agents (Undesirable Practices) (No 2) Order 1991. As far as clients are concerned, the majority of complaints centred on the fear that there might be a conflict of interest and, while the general law of agency could offer at least some protection under the doctrine of 'secret profit',[1] it is clear that by no means all services offered to purchasers would be caught in this way.

The Government's response to this problem lies in regulation 2 of the 1991 Regulations, which requires estate agents to inform their clients (at the same time that they are informed of the agents' terms of business) as to certain services which are to be offered to prospective purchasers. The services in question, which do not include any which are offered free of charge, are defined as 'any services to a prospective

---

1   See pp 70–71.

purchaser ... which are such as would ordinarily be made available to a prospective purchaser in connection with his acquisition of an interest in land or his use or enjoyment of it'. The regulation goes on to list by way of example 'the provision to that purchaser of banking and insurance services and financial assistance and securing the disposal for that purchaser of an interest in land if that disposal is one which has to be made in order for him to be able to make the acquisition he is proposing or is one which is a result of that acquisition'.

This wide definition (which, it is important to emphasise, is in no way limited to the examples given) will clearly include such matters as assisting a prospective purchaser in obtaining a mortgage or a required life insurance policy, or introducing the purchaser to a surveyor or a removal firm. Further, however obvious it might appear to clients who realise that estate agency is a business, an agent must solemnly inform the client that, if a prospective purchaser has a house to sell, the agent will offer to market it. However, the definition does not cover the case of an agent who accepts a bribe from a purchaser, in return for persuading the vendor to accept his offer; such 'services' are not 'ordinarily made available' to a prospective purchaser.[1]

The obligation of an estate agent under regulation 2 arises in three situations. First, and simplest, there is the case where the estate agent (which, it should be remembered, may be a partnership or a company, as well as an individual) itself intends to offer relevant services. Secondly, the services may be offered by what the Regulations call a 'connected person'. The meaning of 'connected person' for this purpose picks up certain definitions in the Estate Agents Act itself, where they are used in setting out the scope of the agent's duty to disclose any 'personal interest'. These terms will accordingly be discussed in detail in relation to section 21;[2] for the moment it will suffice to point out that 'connected persons' include an estate agent's employer, employee, principal or agent, and also any 'associate' of either the agent or of the agent's employer, employee, principal or agent. As for 'associates', these include both business associates such as partners (though not, rather oddly, co-directors of a company) and personal associates such as spouses and a wide circle of relatives. What is more, where an estate agency consists of a partnership or company, there are detailed provisions under which it may be regarded as the 'associate' of an organisation of equivalent type.

---

1   R v *Director-General of Fair Trading, ex p Benhams Ltd* [2001] 1 EGLR 21.
2   See p 299.

The third situation in which a client must be told about 'services' is where these are to be offered by an independent third party, but where the estate agent or a 'connected person' will derive a financial benefit from this. This would apply, for example, where an estate agent regularly introduces prospective purchasers to a particular mortgage or insurance broker and receives a commission from that broker for the introduction.

Although the duty created by this regulation is wide-ranging, it should be emphasised that the estate agent is not obliged to give details of the services in question; the regulation states 'information ... as to the services', which appears to require no more than a list, and certainly does not compel the agent to reveal the amount of any financial benefit which will accrue. Still less does the regulation demand that the agent should obtain the client's permission to offer such services, although a client who objects strongly enough can of course withdraw the agent's instructions altogether.

Time and manner of compliance

Somewhat surprisingly, section 18 as originally enacted did not specify any particular manner in which the required information was to be given to clients. In consequence, while any sensible agent would give the information in writing (so as to be able to prove compliance, if called upon to do so), what mattered in law was simply whether or not the client had been informed.[1] However, regulation 4 of the Estate Agents (Provision of Information) Regulations 1991, issued under section 18(4)(b) of the 1979 Act, provides that all relevant information within section 18 shall be given by the estate agent in writing.

The manner in which information is to be given is clear enough; the time at which it must be given is less so. Section 18(1) as enacted required this to be done 'before any person (in this section referred to as 'the client') enters into a contract with another (in this section referred to as 'the agent') under which the agent will engage in estate agency work on behalf of the client'.[2] There were considerable difficulties of analysis in relating this to the generally held view of estate agency as a

---

1   See *Fiesta Girl of London Ltd* v *Network Agencies* [1992] 2 EGLR 28.
2   The provision appears not to bite unless a contract is entered into. Hence, an agent who attempts to entice another agent's client cannot be guilty of a breach of section 18 unless the attempt succeeds!

unilateral or 'if' contract;[1] however, these problems can now for practical purposes be ignored, since the time at which section 18 must be complied with has been spelled out more precisely in the 1991 Regulations. According to regulation 3(1), the relevant time is 'when communication commences between the estate agent and the client or as soon as is reasonably practicable thereafter provided it is a time before the client is committed to any liability towards the estate agent'.

It is important to appreciate that this provision does not mean that agents can delay in giving the necessary information to their clients so long as they give it before the client is committed to any liability for fees or expenses. The basic requirement is to give the information as soon as communication commences, which suggests for example that an agent visiting a potential client to measure up the property which is to be marketed should take along a copy of the firm's terms of business; the agent can thus ensure that this is given to the client before any 'estate agency work' is carried out. This is not to say that a letter 'confirming instructions', if sent out promptly, would fail to satisfy the regulations; however, there is always the risk that such a letter might not be regarded as having been sent 'as soon as is reasonably practicable' and, in any event, it can never be in time if any charges have already been incurred.

Sanctions for non-compliance

Any breach by an estate agent of section 18 (including those matters governed by the 1991 Regulations) may have two legal consequences. First, as with the other specific duties which are imposed by the Act, such a breach may operate to 'trigger' the enforcement powers of the Office of Fair Trading. Second, it is provided by section 18(5) that, unless the information has been given at the correct time and in the correct manner, the contract is unenforceable, either by legal action or by the exercise of a lien over the client's money, without a court order.[2] Where such an order is sought, the court is given a discretion by section 18(6) either to dismiss the agent's claim altogether, where this is justified by prejudice caused to the client by the agent's failure to comply with his obligation and the degree of culpability for the failure', or to reduce the amount payable so as to compensate the client

---

1   See pp 43–46.
2   But compliance with the Act does not necessarily render the agent's commission claim legally enforceable. In particular, there are doubts as to the status of 'confirming letters': see pp 141–144.

for any prejudice suffered (presumably in cases where the agent's failure cannot be described as 'culpable').

In the Scottish case of *Solicitors Estate Agency (Glasgow) Ltd* v *MacIver*,[1] a client complained that, by not disclosing the discount of 18 per cent which they received on the charge for advertising the client's property in a local newspaper, estate agents were in breach of either section 18(2)(a) or section 18(2)(d) or both (depending on whether or not the money which thereby accrued to the agents was to be regarded as part of their 'remuneration'). This complaint was upheld and, having taken into account both the agents' 'culpability' and the client's 'prejudice', the court ruled that the commission payable should be reduced by one-half.

Subsequent attempts by clients to use section 18 as a ground for refusing to pay fees have proved much less successful. In *Connells Estate Agents* v *Begej*,[2] for example, where the agents claimed commission for bringing about a part-exchange of residential properties, the client argued that their terms of business failed to make any provision for this. However, the Court of Appeal held that the agents had complied with section 18; their agency terms stated that commission would be payable on completion of the sale, and a part-exchange constituted a 'sale' for this purpose.[3]

Even where a breach of section 18 is established, it does not automatically follow that the agent loses the entitlement to fees. In *Benhams Ltd* v *Kythira Investments Ltd*,[4] for example, the claimant agents had acted regularly for the defendant property development company in both the sale and, less frequently, the acquisition, of residential properties. On one occasion, the agents received oral instructions to act and never set out their terms in writing. It was nevertheless held that, since the clients were well aware of the agents' normal terms (which had applied in all their previous dealings), they had suffered no 'prejudice' as a result of the lack of written notice; it would therefore not be just to deprive the agents of their commission.[5]

---

1   1993 SLT 23.
2   [1993] 2 EGLR 35.
3   See also *MSM Consulting Ltd* v *United Republic of Tanzania* [2009] EWHC 121 (QB), where it was held that a purchasing agent's entitlement to commission on a transaction taking place after termination of the agency had been satisfactorily explained.
4   [2004] EWHC 2973 (QB).
5   A similar decision was reached in *G & S Properties* v *Henderson* (6 January 1999, Ct Sess (OH)), where joint sole agents were successful in an action for their fees, despite not having provided an explanation of the term 'sole selling rights'.

Other duties to clients

As far as estate agents' duties to their clients are concerned, the 1979 Act itself has little to offer other than the information-giving requirement of section 18. However, the 1991 Regulations and Orders are rather more demanding in this respect, laying down a number of duties which continue throughout the agency relationship. These may now be considered.

Variation of terms

As noted above, section 18 of the 1979 Act and the Estate Agents (Provision of Information) Regulations 1991 impose detailed obligations upon estate agents to give their clients written notice of a number of matters at the commencement of the agency relationship. Of course, it is perfectly possible for an estate agent and a client to agree at some point that the terms of their contract shall be varied. If this happens, section 18(3) requires the agent to notify the client of the new terms; the 1991 Regulations require the updated information to be given in writing at 'the time when, or as soon as is reasonably practicable after', the changes are agreed by the parties.

Services requested by purchasers

We noted above the obligation imposed upon an estate agent by the Provision of Information Regulations, to inform the client of any 'services' which the agent intends to offer to prospective purchasers. It is further provided by the Estate Agents (Undesirable Practices) (No 2) Order 1991 that, where an estate agent has introduced a prospective purchaser to the client and that purchaser has made an offer,[1] the agent must also inform the client 'promptly and in writing' of any services[2] which that purchaser requests from the agent, or from a 'connected person',[3] or from anyone else in circumstances where the agent or a connected person will derive a financial benefit. This obligation applies to all such requests, except those which are refused outright,

---

1   'Offer' for this purpose is expressly defined to include a conditional offer.
2   For the definition of services, see *R* v *Director-General of Fair Trading, ex p Benhams Ltd* [2001] 1 EGLR 21: p 291.
3   See p 299.

which are received at any time before contracts for the sale or lease of the property are exchanged.

It is important to emphasise that this obligation is completely independent of the one discussed earlier. Even if the agent has already warned the client that, for example, the firm will offer to arrange mortgages for prospective purchasers, the client must be told when a purchaser actually requests the firm to do so. Furthermore, where an agent has not informed the client at the outset of a particular service (because at that time the agent had no intention of offering it), the agent must nonetheless report any request from a prospective purchaser unless this is immediately refused.

Notification of offers received

The Estate Agents (Undesirable Practices) (No 2) Order 1991 seeks to ensure that estate agents are completely open with their clients over offers received from prospective purchasers. To this end, Schedule 3 para 2 defines as an undesirable practice any 'failure by an estate agent to forward to his client ... accurate details ... of any offer the estate agent has received from a prospective purchaser in respect of an interest in the land'. These details are to be forwarded 'in writing' and may be sent 'by hand, post or fax at the address or to the number given by the client to the estate agent'. Whatever method is adopted, the information must be sent 'promptly', which is defined to mean 'within as short a period as is reasonably practicable in the circumstances, from the moment when what is to be done can reasonably be done'.

As to precisely what information must be sent to the client, the 1991 Order is not explicit. There is no explanation of what constitutes 'details', although it may be suggested that these should include at least the amount of the offer, the identity of the person making it, and perhaps the form in which it is received (whether or not it is in writing). Moreover, it was suggested in *R v Director-General of Fair Trading, ex p Benhams Ltd*[1] that an agent who received a bribe from a prospective purchaser, in return for persuading the vendor to accept his offer, was guilty of a breach of this provision in not informing the vendor about the bribe!

It should be noted that the agent's duty applies as much to conditional offers as to unconditional ones, which suggests that the

---

1   [2001] 1 EGLR 21.

agent should surely identify any conditions which are attached to an offer (eg whether or not it is 'subject to contract'). However, an important exception to this 'undesirable practice' lies in the fact that it does not extend to categories of offer, or categories of 'details', which the client has indicated in writing need not be forwarded. For example, a client might agree that the agent need not pass on offers which are less than a certain percentage of the asking price.

Disclosure of personal interest in purchase

Section 21 of the Estate Agents Act imposes a duty of disclosure in respect of certain interests which either the agent or a person connected with the agent has in property to be disposed of or acquired. The main thrust of that provision concerns disclosure to third parties with whom the agent 'enters into negotiations'; indeed the extent if any to which it can compel disclosure to the client is highly debateable.

This potential loophole has now been directly addressed by Schedule 1 para 2 to the Undesirable Practices Order, which requires an estate agent to disclose to the client 'promptly and in writing' that the agent 'has, or is seeking to acquire, a beneficial interest in the land or in the proceeds of sale of any interest in the land'. A similar obligation arises where the agent 'knows that any connected person[1] has, or is seeking to acquire' such an interest.

The duty of disclosure to the client which is imposed by this paragraph appears to arise in four different situations:

- Where an agent acting for a vendor or landlord already has a 'beneficial interest' in the property. This would arise, for instance, where a tenant of property instructs an estate agent to market the leasehold interest, and the agent has an existing interest (direct or through a 'connected person') in the freehold.

- Where an agent acting for a vendor or landlord wishes to acquire a beneficial interest in the property.

- Where an agent acting for a client who is seeking property to rent or buy introduces property in which the agent or a 'connected person' has a beneficial interest.

---

1   The meaning of 'connected person' is discussed at p 299.

- Where an agent, acting on behalf of a client who is seeking property to rent or buy, introduces a property to the client but then decides to acquire that property personally or for a connected person (ie the agent intends to compete with the client for the property).

Duties to third parties

A combination of the 1979 Act and the Estate Agents (Undesirable Practices) (No 2) Order 1991 means that, in addition to their clients, estate agents also owe certain obligations to third parties with whom they deal in the course of business. Three matters are of particular importance in this respect.

Disclosure of personal interest

Once an estate agent accepts instructions to act for a client, a fiduciary relationship comes into existence between them, as a result of which an equitable duty of loyalty is imposed upon the agent.[1] This duty manifests itself in various ways, but these are all related to the principle that an agent's personal interests should not be allowed to come into conflict with the interests of the client. If a potential conflict arises, it is the agent's duty to make full disclosure to the client, so that the latter can make an informed decision on what action to take. Failure to make such disclosure may render the agent liable to pay damages, disgorge any personal profit which has been made and forfeit the agreed commission.

The equitable duty of loyalty arises out of the agency relationship; there is no corresponding obligation owed to a third party with whom the agent deals. As a result, it appears that an auctioneer is quite entitled to sell his or her own property, without revealing its ownership to bidders.[2] Further, although there is no authority on the point, it seems likely that the same would be true of an estate agent. However, such cases are now subject to a statutory duty of disclosure by virtue of section 21 of the Estate Agents Act.

The first point to be noted about section 21 is that it is not limited in its application to the carrying out of 'estate agency work' as defined

---

1   See p 62.
2   *Flint* v *Woodin* (1852) 9 Hare 618: p 78.

by section 1 of the Act. The opening words of section 21(1): 'A person who is engaged in estate agency work' might, if taken out of context, mean either someone generally engaged in estate agency work (ie someone whom the layman would describe as an 'estate agent') or someone so engaged in the relevant transaction. However, section 21(3) shows conclusively that the former interpretation is the correct one by providing that subsections 1 and 2 apply 'where an estate agent is negotiating on his own behalf as well as where he is negotiating in the course of estate agency work'. As a result, this statutory duty of disclosure applies, not only when an estate agent acts pursuant to instructions received from a client, but also when the agent decides to enter the marketplace in a personal capacity.

Before we turn to consider the actual obligations which are imposed by section 21, some attention must be paid to what that section regards as a 'personal interest'. In one sense, this concept is less widely drawn by statute than in equity, for it is limited to interests in the land itself. In other respects, however, the scope of the section far surpasses any duty created by the courts.

Section 21(5) provides:

> for the purposes of this section, an estate agent has a personal interest in land if —
>
> (a) he has a beneficial interest in the land or in the proceeds of sale of any interest in it; or
> (b) he knows or might reasonably be expected to know that any of the following persons has such a beneficial interest, namely —
>
>   (i) his employer or principal, or
>   (ii) any employee or agent of his, or
>   (iii) any associate of his or of any person mentioned in sub-paragraphs (i) and (ii) above.

The overall width of this definition is enormous. In the first place, what we may term an estate agent's 'direct' personal interests in land will include any interest whatsoever[1] in either the land itself or in the proceeds of sale of any interest in it. Secondly, the estate agent will have what we may term an 'indirect' personal interest in land if any member of a wide class of persons associated with the agent has a 'direct' personal interest and the agent is or should be aware of that fact. This

---

1   Not just the freehold or lease included in s 2 but any legal or equitable interest.

class covers not only the estate agent's partner, employer,[1] employee principal and agent, but also his and their 'associates', who are defined by section 32 as including spouses and an extensive circle of relatives.

In an effort to provide an exhaustive definition of 'personal interest', the Act deals in sections 31 and 32 with the common situation where an estate agency practice is carried on, not by an individual, but by a partnership, company or unincorporated association. In each case, certain natural persons are named as 'business associates' of the organisation, which is then under a duty to disclose any direct personal interest of those persons, their spouses and relatives. Further, complex provisions govern the possibility that two partnerships, two companies or two unincorporated associations may be 'associates' where certain natural persons are common to both organisations. In any such case, one associate would be under a duty to disclose any direct personal interest of the other. Interestingly, however, no provision is made for the possibility of one type of organisation, such as a limited company, being the 'associate' of another type, such as a partnership. Thus, for example, if X, a partner in a firm of estate agents, is also a director of a development company which markets its houses through X's firm, X as director will be deemed to have a personal interest in the houses, and thus a duty of disclosure. However, the partnership as a whole has no such interest or duty. As a result, so long as X does not personally conduct negotiations with any third party but leaves everything to the other partners, there is no breach by anybody of section 21. If, however, X does contravene the section, it must be remembered that the partnership (and the other individual partners) may also be responsible for that contravention as 'business associates' under section 3(3)(c), provided that they are guilty of 'connivance or consent'.

On turning to the actual obligations as to disclosure which are created by section 21, it may be seen that the section attempts to deal with two different situations. First, section 21(1) provides that an estate agent who has a personal interest in land shall not 'enter into negotiations with any person with respect to the acquisition or disposal by that person of any interest in that land until the estate agent has disclosed to that person the nature and extent of his personal interest in it'. At its simplest, this covers the case where an estate agent is marketing his or her own house; here the truth about ownership

---

1   Note, however, that the agent who is acting on behalf of his own employer may in any case be outside the scope of the Act: see p 278.

must be disclosed to any prospective purchaser before negotiations commence. This, however, is only the tip of the iceberg; the agent's duty of disclosure applies equally where the personal interest is 'indirect' (as described above) and also where that interest in the land is not the one which is being sold. For example, an estate agent instructed to sell the freehold reversion on business premises would come within section 21(1) if the agent's nephew happened to be the sub-lessee of those premises, or if the ex-wife of one of the agent's partners held a mortgage over them (provided in each case that the agent knew or ought to know the relevant facts).

The examples so far given all relate to an estate agent who is instructed to sell property, but the wording of section 21(1) clearly also covers the converse case. It may be unusual for an agent to be negotiating with a third party over land in which they both hold a personal interest without this being obvious but, given the wide definition of personal interest in this section, it is by no means impossible. Suppose, for example, that an estate agent is approached by a client who is seeking leasehold shop premises to rent. Suppose also that an 'associate' of that agent is the landlord of a suitable shop. By virtue of the provision under discussion, the agent may not begin to negotiate with the present tenant for an assignment of the lease until the nature and extent of the agent's personal interest in the property have been disclosed to that tenant.

The Estate Agents Act itself does not specify precisely either how or when agent's duty of disclosure is to be fulfilled, beyond stating that this is to be done before 'negotiations' commence. However, this matter has been effectively fleshed out by Schedule 1 to the Estate Agents (Undesirable Practices) (No 2) Order 1991, which provides that any required disclosure is to be made 'promptly and in writing'. As to what is meant by 'promptly', the Order defines this as 'within as short a period as is reasonably practicable in the circumstances, from the moment when what is to be done can reasonably be done'.

If section 21(1) appears nightmarishly complex, what is to be made of section 21(2)? This deals with cases where the estate agent has no initial personal interest in the land in question, but where, as a result of a proposed transaction or series of transactions, such an interest will be acquired.[1] One might regard the case of an estate agent

---

1 Note that disclosure in this instance need not be made 'promptly and in writing', since the Estate Agents (Undesirable Practices) (No 2) Order 1991 does not apply.

buying a client's property through an undisclosed 'associate' as the obvious abuse at which this provision is aimed but, on closer examination, section 21(2) does not appear to catch this situation. Once again, the duty of disclosure is owed to any person with whom the agent 'enters into negotiations', and it seems to place an unjustified strain upon the ordinary meaning of these words to suggest that they include the agent's own client. What section 21(2) covers is the case where an estate agent seems to be buying on behalf of an independent client, whereas in reality the agent intends to act personally or for an associate. Even so, it is by no means certain that the common case of an agent bidding at auction would be included, since such conduct might well not amount to 'entering into negotiations'.

Interestingly, a situation which appears to fall outside both section 21 of the Act and the Undesirable Practices Order, despite having been singled out for adverse comment by the Department of Trade and Industry, is where an estate agent acting on behalf of a vendor wishes to make a personal bid for the property (or to sell it to a connected person). The agent's intention must in such a case be disclosed to the client;[1] however, it seems that there is no obligation to disclose to prospective purchasers the fact that the agent intends to compete with them for the property.

All in all, it cannot be said that section 21 is happily drafted. In particular, it seems likely that, because the definitions are so widely drawn, many agents will find themselves in breach of its terms in cases where there is not the slightest suspicion of sharp practice. The sole consolation which can be offered is that there are no criminal or civil sanctions for a breach of these provisions;[2] the only effect is that a breach may be used to trigger the enforcement powers of the Office of Fair Trading, and one may hope that the OFT is unlikely to take any action in respect of trivial or inadvertent breaches.

Discrimination over services

One matter which was identified by the Office of Fair Trading in the late 1980s as giving cause for concern related to those estate agents who, while acting for a vendor, sought directly or indirectly to profit from the provision of services (financial or other) to prospective

---

1   See p 297.
2   S 21(6).

purchasers. It was felt that such practices, which had naturally increased considerably with the move by financial institutions into the estate agency field, could have adverse effects on both purchasers and clients. Unless restrained, an unethical agent might well be tempted for example to favour a prospective purchaser who would require a mortgage from the agent's linked financial services division over one who had already arranged the necessary finance. This would obviously be detrimental to the rejected purchaser; it could also of course work to the disadvantage the client, who might thereby be deprived of a higher offer for the property. True, the client (if able to prove what had occurred) could take action against the agent for any loss suffered and could also refuse to pay commission;[1] however, the enforcement authorities felt that the situation called for legislative intervention.

We have already noted[2] those provisions in the 1991 Orders and Regulations which require an estate agent to inform (and keep informed) the client as to services which are provided directly or indirectly to prospective purchasers. However, the legislation goes further by defining as an undesirable practice any 'discrimination against a prospective purchaser by an estate agent on the grounds that that purchaser will not be, or is unlikely to be, accepting services'.[3]

While 'services' for the purpose of this provision bears the meaning which we have discussed earlier, 'discrimination' is not further defined in the Order. However, other statutory provisions in the fields of sexual and racial equality define discrimination in terms of treating a person less favourably than one treats or would treat others in similar circumstances, and there seems no reason to doubt that such ideas would also be applied in the estate agency context. It would surely be discrimination, for example, to seek to persuade the client to prefer one applicant to another,[4] to delay in passing on an offer from an unfavoured purchaser,[5] or to send out details of suitable properties only to those applicants identified as likely to require services.

---

1   See *Henry Smith & Son* v *Muskett* [1978] 1 EGLR 13: p 68.
2   See pp 290, 295.
3   Estate Agents (Undesirable Practices) (No 2) Order 1991, Schedule 2 para 1.
4   As occurred in *Henry Smith & Son* v *Muskett* [1978] 1 EGLR 13.
5   This would of course also be an undesirable practice under Schedule 3 para 2.

Misrepresentation of offers[1]

An estate agent acting on behalf of a vendor should naturally try to obtain the best possible price for the client and, to that end, to emphasise the attractions and desirability of the property in question. Such conduct is only to be expected, although a feeling that agents have on occasion been somewhat over-enthusiastic in this respect is what has led to the passing of the Property Misdescriptions Act 1991. However, one type of misdescription which is not covered by that Act is any false or misleading statement as to the state of competition for a particular property. This is because claims designed to play off one prospective purchaser against another (for example by suggesting that the other has made a higher offer or is a 'cash buyer') do not appear on the list of 'prescribed matters' which define the scope of the 1991 Act.

This particular matter is addressed by Schedule 3 para 1 to the Estate Agents (Undesirable Practices) (No 2) Order 1991, which provides that it is an undesirable practice for an estate agent to make any misrepresentation '(a) as to the existence of, or details relating to, any offer[2] for the interest in the land; or (b) as to the existence or status of any prospective purchaser of an interest in the land'. This applies to all false or misleading statements, whether made in writing or orally; however, an agent is only in breach where the offending statement is made 'knowingly or recklessly'.

Pretending to one prospective purchaser that a rival has made a higher offer for the property might be regarded by cynics as the equivalent in private treaty sales of an auctioneer's taking of bids 'off the wall'. While that practice is clearly illegal,[3] an auctioneer may quite lawfully bid on behalf of the vendor provided that the right to do so has been properly reserved in the particulars or conditions of sale.[4] The 1991 Order accordingly provides that an auctioneer who bids on behalf of the vendor in accordance with the 1867 Act is not to be regarded as guilty of an 'undesirable practice'.

---

1   Although the main target of this provision is the agent who misleads potential purchasers, it can also apply to one who seeks to mislead the client about an offer: see *R* v *Director-General of Fair Trading, ex p Benhams Ltd* [2001] 1 EGLR 21: p 296.

2   Including a conditional offer.

3   See pp 191–193.

4   Sale of Land by Auction Act 1867, s 6.

Duty to keep records

In order to make it easier for enforcement authorities to check whether an estate agent has complied with the various duties described above, section 21A of the Estate Agents Act[1] imposes an obligation to keep certain prescribed records for a period of at least six years. Such questions as how and where these records are to be kept are to be provided for by regulations; however, at the time of writing no regulations had been made and, indeed, section 21A had not been brought into force.

Restrictions on the right to practise

It has already been noted that, unlike previous attempts to legislate in the field of estate agency, the 1979 Act does not compel would-be practitioners to obtain a licence before setting up in business. The general philosophy is that there should be no restrictions upon the right to practise estate agency. There are, however, three exceptions to this general principle; two are no more than one would expect, but the third is more controversial.

Prohibition orders

As will shortly be seen, the Office of Fair Trading is empowered in certain circumstances to make an order prohibiting a person from engaging in estate agency work, on the ground that that person is unfit to do so. Once such an order is made, it must be obeyed, on pain of criminal sanctions.[2]

Bankruptcy

Section 23(1) provides:

> An individual who is adjudged bankrupt after the day appointed for the coming into force of this section or, in Scotland, whose estate is

---

1 Inserted by the Consumers, Estate Agents and Redress Act 2007, s 54.
2 An unlimited fine following conviction on indictment; a fine of up to £5,000 on summary conviction.

sequestrated after that day shall not engage in estate agency work of any description except as an employee of another person.[1]

This prohibition, which ceases to have effect when the bankrupt is discharged, is backed by criminal sanctions.[2] Interestingly, however, a criminal conviction in such circumstances is not a ground on which the Office of Fair Trading may make an order banning the agent from practice; as a result, the bankrupt agent could still continue to practise as an employee of someone else.

Minimum standards of competence

Somewhat out of place in an Act which firmly eschews any positive system of licensing is section 22(1), which enables regulations to be made prescribing minimum standards of competence for those engaged in estate agency work. This provision, though at first regarded as highly controversial, now commands the support of the major estate agency bodies;[3] however, it has not been brought into force and the present Government has repeatedly stated that there is no intention to implement it.

If regulations are ever made under section 22, they must 'prescribe a degree of practical experience which is to be taken as evidence of competence' and, in addition, they may provide for alternative qualifications. As to what these alternatives might be, section 22(2) provides that regulations may:

(a) prescribe professional or academic qualifications which shall also be taken to be evidence of competence;

(b) designate any body of persons as a body which may itself specify professional qualifications the holding of which is to be taken as evidence of competence;

(c) make provision for and in connection with the establishment of a body having power to examine and inquire into the competence of persons engaged or professing to engage in estate agency work; and

---

1   A possible loophole, by which a bankrupt might form a company and work for it under a contract of employment, is closed by s 23(3).

2   An unlimited fine on conviction on indictment; on summary conviction, a fine of (present) maximum of £5,000.

3   Following a U-turn by the Royal Institution of Chartered Surveyors, which was originally concerned that 'official recognition' might be given to those holding lesser qualifications.

(d)   delegate to a body established as mentioned in paragraph (c) above powers of the Secretary of State with respect to the matters referred to in paragraph (a) above.

Of these provisions, paragraph (a) is straightforward, while paragraph (b) assumes that the existing professional bodies with their own examination systems would be designated to set the standards of competence for their own members, so that existing arrangements would continue largely undisturbed. As for paragraph (c), what is envisaged is a body representative of all persons engaged in estate agency work, whether or not holding professional qualifications, to test the competence of those not wishing to join the designated bodies, to prescribe standards for them, and to provide an alternative to the arrangements which are institutionalised by the existing professional bodies. The composition of this body is to be determined by the regulations, but it seems clear that, at least at the outset, the knowledge and expertise of members of the existing professional bodies would be heavily relied upon.

If section 22 is ever brought into force, any person who engages in estate agency work 'on his own account' will, unless he or she satisfies the required standard of competence, be guilty of a criminal offence.[1] Further, statutory regulations will provide what proportion of partners must satisfy the minimum standard in order to render the partnership itself 'competent', and similar rules will apply to any 'body corporate or unincorporated association' which engages in estate agency work.

Enforcement of the Act

Responsibility for the enforcement of the Estate Agents Act is given by section 26(1) to both the Office of Fair Trading (OFT) and local weights and measures authorities.[2] The latter were originally restricted to the investigation and prosecution of specific criminal offences under the Act; however, the present version of section 11[3] now extends their powers to all those duties of estate agents described earlier in this

---

1   Carrying, on conviction on indictment, an unlimited fine; on summary conviction, the maximum fine is at present £5,000.

2   In Northern Ireland, the functions of the latter are given to the Department of Commerce.

3   As amended by the Consumers, Estate Agents and Redress Act 2007, s 57.

chapter. Notwithstanding this division of responsibility, however, local weights and measures authorities may be called upon to report to the OFT on the carrying out of their functions under the Act.

Negative licensing

The overall supervision and enforcement of the Act is made the responsibility of the OFT. However, unlike the Consumer Credit Act 1974, which the OFT also administers, the Estate Agents Act does not impose any positive system of licensing or registration. With very few exceptions,[1] any person may practise as an estate agent until proved unfit to do so. The scheme, therefore, consists of a kind of 'negative licensing', and the question of a particular agent's fitness or unfitness is a matter for the OFT which may, if the requirements of the Act are satisfied, make various orders, including one which prohibits the agent altogether from continuing to practise.

The taking away of a person's livelihood is of course a drastic form of punishment and, not surprisingly, the OFT's powers in this area are hedged about by numerous restrictions, both procedural and substantive.[2] Not only must the OFT be satisfied that the agent is generally unfit; there must also have occurred one of a number of specified 'trigger' events, such as a criminal conviction, before these powers come into play.

We shall shortly consider the various 'triggers' and the procedures (including the question of appeals) which govern their exercise. First, however, we must look at the important matter of the exact range of orders which the OFT is empowered to issue.

Types of order

The simplest and most drastic form of order which may be made under the Estate Agents Act 1979 is one which prohibits a person[3] from doing any estate agency work at all. If, however, the OFT feels that the 'unfitness' relates to only one aspect of estate agency work, the order

---

1   See p 305.
2   It has been held that, where a person's livelihood depends upon the decision of a tribunal, the strict test of proof 'beyond reasonable doubt' is to be applied: see *R* v *Milk Marketing Board, ex p Austin* [1983] *The Times*, March 21.
3   Or an estate agency partnership or company as a whole: see p 309–310.

may instead prohibit the agent 'from doing estate agency work of a description specified in the order'.[1] A limited order of this kind might be thought appropriate, for example, where an agent's careless accounting procedures, while not reflecting upon his or her general level of honesty and competence, nevertheless make the agent unfit to hold clients' money. In such a case, the agent might simply be banned from accepting clients' money.[2]

It is further provided[3] that any order, whether general or specific, may be limited in scope to a particular part of the United Kingdom, a useful provision where, for example, complaints of malpractice are restricted to one branch of a larger organisation. There is, however, no power to limit the time for which an order is to operate; any ban takes effect (subject to the possibility of revocation following the agent's application)[4] as a life sentence.

The severity of a prohibition order may be tempered under section 4, which empowers the OFT in certain cases to issue a warning to the estate agent that, if he or she persists in infringing the law, the OFT will make a declaration of unfitness to practise under section 3. It is vital to appreciate that this warning power does not exist where the grounds for an order under section 3 consist of a criminal conviction (section 3(1)(a)) or either racial or sex discrimination (section 3(1)(b)). Further, it should be emphasised that a warning order can only be made where the OFT is satisfied that the agent's general 'unfitness' will justify an order under section 3 if the warning order is not heeded.

A warning order made under section 4 must indicate the type of section 3 order (general or limited) which will be made against the agent if there is a further infringement. If such an infringement does occur,[5] the OFT may automatically proceed to make an order under section 3; the agent's failure to comply with the warning is, by section 4(3), to be treated as conclusive evidence of unfitness to practice.

It should be noted that any prohibition or warning order under the Estate Agents Act is made against a 'person' which, according to

---

1   S 3(2).
2   As to money which is already in the agent's possession when the order comes into force, the OFT is empowered to appoint a new trustee in substitution for the banned agent: see s 13(4).
3   By section 3(5).
4   See p 319.
5   If the warning order is based on a breach of the Act itself (ie sections 15 and 18 to 21), a further breach of any of those sections may trigger a ban. If it is based on an 'undesirable practice', however, it must be the same practice which recurs.

the Interpretation Act 1978, 'includes a body of persons corporate or unincorporate'. As a result, it seems clear that an order may be made against a partnership or a limited company, as well as an individual estate agent. Indeed, the possibility of dealing en bloc with a partnership is expressly envisaged by section 5(2), which provides that the order may also have effect against some or all of the partners individually. Somewhat oddly, however, there is no similar provision in relation to limited companies.

Trigger events

The grounds upon which an order may be made by the OFT are set out in section 3(1). This requires the OFT to be satisfied that the agent:

(a) has committed –

  (i) an offence involving fraud or other dishonesty or violence, or
  (ii) an offence under any provision of this Act, other than section 10(6), section 22(3) or section 23(4), or
  (iii) any other offence which, at the time it was committed, was specified for the purpose of this section by an order made by the Secretary of State; or

(b) has committed discrimination in the course of estate agency work; or

(ba) has failed to comply with an undertaking accepted from him under section 217, 218 or 219 of the Enterprise Act 2002 and given in relation to estate agency work; or

(bb) has failed to comply with an enforcement order under section 217 of the Enterprise Act 2002 which was made against him in relation to estate agency work; or

(c) has failed to comply with any obligation imposed on him under any of sections 15 and 18 to 21A below; or

(ca) has engaged in estate agency work in relation to residential property in breach of the duty imposed by an order under section 23A(1) below; or

(cb) has failed to comply with any requirement imposed on him under section 9(1) or 11(1A)(b) below; or

(d)   has engaged in a practice[1] which, in relation to estate agency work, has been declared undesirable by an order made by the Secretary of State.

The categories of events which trigger the OFT's powers may now be considered.

Criminal offences

Section 3(1)(a) lists three groups of criminal offences, commission of which will justify the OFT in making an order against the agent.[2] In relation to all of these it should be noted that a conviction which is 'spent' by virtue of the Rehabilitation of Offenders Act 1974 'or any corresponding enactment for the time being in force in Northern Ireland' is to be disregarded.[3]

The first group of offences, those involving fraud, dishonesty or violence, is identical to that which would justify the OFT in refusing to grant a licence under the Consumer Credit Act 1974. It is important to note that commission of an offence within this group is a 'trigger', whether or not the offence took place in the course of estate agency work. Indeed, the OFT has used this provision to issue orders against persons who were not estate agents at all, but who it was thought might start an estate agency business (and who would be highly unsuitable if they did).

In *Antonelli* v *Secretary of State for Trade and Industry*,[4] the applicant was convicted in 1973 by an American court of 'burning real estate other than a dwelling house', an offence for which he was sentenced to between two and a half years and 10 years imprisonment. This conviction was subsequently relied upon by the OFT as a ground for a order under section 3. The applicant's challenge to the order was rejected by both a high court judge and the Court of

---

1   It has been held that a single act may be a sufficient 'practice' for this purpose: there is no requirement of repetition or continuity: see *R* v *Director-General of Fair Trading, ex p Benhams Ltd* [2001] 1 EGLR 21: p 296.

2   As first enacted, these criminal offences became triggers only when a person was actually convicted. The amended wording (which enables the OFT to decide that an offence has been committed, even though there has been no prosecution) was introduced by the Consumers, Estate Agents and Redress Act 2007, s 55(2).

3   Schedule 1 para 1.

4   [1998] 1 EGLR 9.

Appeal, which held that this was an 'offence involving violence', albeit violence against property. Moreover, the conviction could be used as a 'trigger', notwithstanding that it was a conviction by a court outside the United Kingdom and that it occurred before the Estate Agents Act came into force.

As to the 15 offences created by the Act itself most of which are concerned either with the handling of clients' money or with obstruction of the enforcement authorities, all but three rank as 'triggers'. Of those which do not, section 10(6) (which makes it illegal to disclose information obtained by authorised officers under the Act) is simply irrelevant to the practice of estate agency. The other two (practising without possessing the necessary qualifications[1] or when bankrupt[2]) are such as to render a banning order superfluous. It should be noted, however, that a criminal conviction for disregarding a previous order of the OFT is a ground for a subsequent order; this might be relevant where, for example, the first order is limited as to type or work or geographical area.

The third group includes any criminal offence which may be specified by statutory instrument. Under this provision, the Estate Agents (Specified Offences) (No 2) Order 1991 lists more than 40 offences arising under 13 statutes. These offences range from the making of unauthorised disclosures by a computer bureau[3] to the impersonation of a licensed conveyancer.[4] The broad impression is that these offences have for the most part been collected from 'consumer protection' statutes, rather than with any real consideration of what it is that estate agents actually do; however, the Estate Agents (Specified Offences) (No 2) (Amendment) Order 1992 is more to the point, in that it adds offences arising under the Property Misdescriptions Act 1991.

The Estate Agents Act contains two provisions which are designed to ensure that, so far as possible, the OFT receives information of any relevant convictions. In the first place, when a local weights and measures authority propose to institute proceedings for an offence under the Act it is their duty to notify the OFT and to supply a summary of the facts upon which the prosecution is to be based.[5] Second, section 9(5) amends section 131 of the Fair Trading Act

---

1   S 22 (not yet in force).
2   S 23.
3   Data Protection Act 1984, s 15.
4   Administration of Justice Act 1985, s 35.
5   S 26(2).

1973 so as to empower courts, both civil and criminal, to notify the OFT of any conviction or judgment which is regarded as relevant to the OFT's functions under either Act.

Discrimination

At first sight, section 3(1)(b) seems to be of enormous scope. This appearance, however, is deceptive; 'discrimination' for the purposes of the Estate Agents Act bears only that meaning which is assigned to it by Schedule 1. In effect, this provides that the question whether or not an estate agent has committed discrimination is not one for the OFT to answer; it is governed by the outcome of previous proceedings under either the Sex Discrimination Act 1975 or the Race Relations Act 1976.[1] As a result, an agent will only be guilty of discrimination under section 3 where a court has made a finding to that effect or where a non-discrimination notice served on the agent by the Equal Opportunities Commission or the Commission for Racial Equality has become final; in either event, the agent will have had an opportunity in judicial or quasi-judicial proceedings to contest the allegations.[2]

To constitute a 'trigger' for the purposes of section 3, the discrimination in question must have been committed 'in the course of estate agency work'. Whether or not it is so committed is a matter for the OFT to decide, but there is an important limitation; an estate agent who discriminates as an employer rather than as an estate agent is not to be dealt with under the Estate Agents Act.

The general treatment of discrimination in the Act runs parallel to that which is given to criminal offences, and this similarity is heightened by two procedural provisions. In the first place, a finding or notice of discrimination under Schedule 1 becomes 'spent' after five years. Second, section 9(6) makes it the duty of the statutory Commissions to notify the OFT of any finding, notice, injunction or order which appears to them to be relevant to the OFT's functions under the Act.

---

1   In relation to Northern Ireland, the reference is to the Sex Discrimination (Northern Ireland) Order 1976 and there is no provision as to racial discrimination.
2   For a case of racial discrimination involving a London firm of estate agents, see *R v Commission for Racial Equality, ex p Cottrell & Rothon* [1980] 3 All ER 265.

Breaches of the Estate Agents Act

Of the specific 'estate agency' duties which are imposed by the 1979 Act, only those which relate to the keeping of client accounts are backed by criminal sanctions. It is, of course, essential that breaches of the other duties should serve as 'triggers' for the enforcement powers of the OFT, and this is achieved by section 3(1)(c), which refers to 'sections 15 and 18 to 21A'.[1] The task of deciding whether an agent has failed to comply with any of these obligations is one for the OFT's own judgment; it is possible, however, for a breach of sections 15, 18 or 20 to lead to a civil action,[2] in which case the court concerned may notify the OFT of its decision under section 131 of the Fair Trading Act 1973.[3]

Undesirable practices

Section 3(1)(d) empowers the Secretary of State to designate as 'undesirable' certain practices which may be found in estate agency work, so as to constitute them as additional 'trigger' events. The six practices so designated by the Estate Agents (Undesirable Practices) (No 2) Order 1991 have been considered earlier in this chapter.

Orders and undertakings under the Enterprise Act 2002

Part 8 of the Enterprise Act 2002 provides a framework for the enforcement by the OFT of certain consumer legislation designed to protect consumers. Briefly, where an act or omission in the course of a business is harmful to 'the collective interests of consumers in the United Kingdom', the OFT (or other enforcement authority, such as a local weights and measures authority) can apply to court for an enforcement order directing the offender not to continue or repeat the offending conduct. However, before applying for such an order, the enforcement authority must 'engage in appropriate consultation' with the offender, with a view to achieving the cessation and non-repetition of the conduct in question without further proceedings. In such circumstances, the

---

1   S 21A (which imposes a duty on the estate agent to keep certain records) is not yet in force: see p 305.
2   The same is potentially true of section 19, but this has not yet been brought into effect.
3   As amended by s 9(5) of the Estate Agents Act.

enforcement authority may accept an undertaking from the person concerned that the conduct will cease or will not be repeated.

Not surprisingly, section 3 of the Estate Agents Act treats both an enforcement order and a breach of an undertaking as 'triggers' for the powers of the OFT under that Act.

Non-membership of redress scheme

As we have seen,[1] any person engaging in estate agency work in relation to residential property is required to be a member of an approved redress scheme. Failure is designated as a 'trigger'.

Failure to co-operate with enforcement authority

As we shall see, the Estate Agents Act confers on authorised officers of an enforcement authority (the OFT and local weights and measures authorities) a range of investigatory powers, including for example that of demanding information (the OFT only) or requiring the production of books, documents or other information stored electronically. Importantly, these powers now extend to the investigation, not only of criminal offences under the Act (a very limited range) but also of any breach of the other duties imposed.

Where a requirement is made under any of these provisions, failure to comply with it is a 'trigger'.

Vicarious liability

An important feature of the last groups of 'trigger' events (that is, all except criminal offences and discrimination) is the extent to which an estate agent may be held responsible for someone else's default. Section 3(3) lays down three separate types of vicarious liability which can apply in this area. In the first place, a person is responsible for anything done by an employee 'in the course of his employment', unless it can be can shown that the employer 'took such steps as were reasonably practicable to prevent the employee from doing that act, or from doing the course of his employment acts of that description'.

---

1   Under s 11A, inserted by the Consumers, Estate Agents and Redress Act 2007, s 58(1).

Second, a person is responsible for any act of an agent (as opposed to an employee) which is expressly or impliedly authorised, a provision which could well lead to problems in cases where an estate agent has sub-instructed.

These two kinds of vicarious responsibility are modelled on those found in the Sex Discrimination Act 1975, section 41 and the Race Relations Act 1976, section 32. By contrast, the third situation covered by section 3(3) creates a wider form of responsibility than one is accustomed to see. This lies in the provision that: 'Anything done by a business associate of a person shall be treated as done by that person as well, unless he can show that the act was done without his connivance or consent.' Given the Act's broad definition of 'business associate',[1] the requirement of 'connivance or consent' seems a necessary safeguard against an unacceptably onerous obligation. However, it should be stressed once again that, like all 'triggers', these events do not compel the OFT to make an order against an estate agent; they merely enable such an order to be made if the agent is considered to be generally unfit to practise.

Proceedings under the Act

The attention of the OFT may become focused on a particular agent or firm for a variety of reasons. There may for example be a complaint from a member of the public (either directly to the OFT or passed on by a trading standards department); a court or tribunal may report a case in which an estate agent has been held guilty of some offence; or a professional body may pass on the result of serious disciplinary proceedings against one of its members. In any of these cases, the OFT may decide that further investigation of the agent is justified.

Investigation by OFT

An investigation by the OFT can be extremely wide-ranging; section 3(2) provides that, in addition to 'trigger' events, the OFT may 'also take account of whether, in the course of estate agency work or any other business activity, that person has engaged in any practice which involves breaches of a duty owed by virtue of any enactment, contract or rule of law and which is material to his fitness to carry on estate

---

1   See s 31.

agency work'. Thus, provided that at least one 'trigger' can be found, an estate agent's unfitness (leading to the loss of livelihood) may ultimately rest upon the conduct of business activities in a very different field.

In carrying out an investigation, the OFT is given far-reaching powers by section 9 of the Act. This empowers the OFT, by simple notice, to 'require any person to furnish to him such information as may be specified or described in the notice or to produce to him any document so specified or described'.[1] Once served, a notice under section 9 must be treated seriously; deliberately giving false or misleading information, or producing a false or misleading document, is a criminal offence. However, a person who refuses or neglects to comply with the notice is no longer at risk of criminal proceedings; instead, the OFT may seek a court order compelling compliance,[2] and may also treat the initial failure to comply as a trigger for its powers under section 3.

The confidentiality of information collected by the OFT under the Estate Agents Act is protected by section 10, which makes it a criminal offence to disclose such information without the consent of the person concerned or, if the information is about a business still in existence, without the consent of the person carrying on that business. However, these provisions do not prevent the disclosure of information for the purpose of criminal investigation or proceedings, civil proceedings under the Act or other consumer protection legislation, or the performance by the OFT of any other statutory functions.

Preliminary proceedings

It is by no means certain that the investigation of an estate agent will lead to formal proceedings by the OFT. However, if the OFT is minded to make a prohibition order under section 3 or a warning order under section 4, the procedures which are set out in Part I of Schedule 2 must be followed.[3] The first step is to notify the person concerned[4] of the

---

1   However, this does not compel any counsel or solicitor to disclose privileged information: s 9(3).

2   Under s 11A, inserted by the Consumers, Estate Agents and Redress Act 2007, s 58(1).

3   The OFT, in the exercise of these adjudicating functions, is governed by the Tribunals and Inquiries Act 1971: see s 24.

4   This initial notice need only be served on a partnership as a whole, even where it is intended that the order shall also take effect against the individual partners. However, all subsequent notices during the proceedings must be served by the OFT on the partners as well.

OFT's proposal. This must be done in some detail; notice of a proposed prohibition order must specify any relevant 'trigger' on which the OFT relies, list other matters which have been taken into account and, if the order is to be based on a previous warning under section 4, make that clear. The notice must give the person affected at least 21 days in which to make written representations to the OFT or, if preferred, to notify the OFT that an oral hearing is required. Where an oral hearing is requested, this is to be arranged by the OFT, and the person affected is to be given at least 21 days notice. The hearing itself is informal,[1] although the estate agent may, if he or she wishes, be represented.[2]

The making of orders

If the OFT decides (having duly taken into account any oral or written representations) to make the order in the form of the original proposal, this may be done. Alternatively, an order may be made in different form, provided that the grounds for this were contained in the original proposal (for example, the OFT may abandon one or more of the original grounds, where others remain). If, however, the OFT wishes to rely on new grounds, the person affected must be given a fresh notice of this proposal, whereupon the whole procedure begins again.

When the OFT decides to make an order, the person or persons affected must be given notice specifying the type of order, the 'trigger' on which it is based, and any other facts which justify it. That person must also be informed of the right to appeal against the order.

As a general rule, an order comes into operation only when there is no further possibility of an appeal against it, either because an appeal has failed or because it is too late for one to be made. If, however, the OFT feels it necessary (for example, on the ground of consumer protection) it may take effect immediately upon the giving of the notice. Once the order is in force, it is a criminal offence under section 3(8) not to comply with it, and a conviction under this provision is itself a 'trigger' event. An order, once made, remains in force until either the conviction or discrimination upon which it is

---

1   Evidence may not be excluded solely on the ground that it would not be admissible in a court of law.

2   By any other person (not necessarily a barrister or solicitor). However, barristers and solicitors have the advantage that communications with them are privileged from disclosure under s 9 or s 11.

based becomes 'spent'[1] or until it is revoked by the OFT on an application made to him under section 6 of the Act.

Appeals and similar proceedings

Once an order is made under either section 3 or section 4, the person affected may try in two different ways to remove its effect. In the first place, it is provided by section 7 that an appeal against the OFT's decision lies to the Secretary of State; if this appeal proves unsuccessful, there may be a further appeal on a point of law to the High Court. There is no right of appeal to the court on questions of fact, but it should be remembered that 'points of law' include the assertion that a verdict was so contrary to all the evidence that no reasonable tribunal could have reached it.

Details of the machinery governing appeals to the Secretary of State are contained, not in the Estate Agents Act itself, but in the Estate Agents (Appeals) Regulations 1981. Briefly, these regulations (which follow closely those applicable to appeals under the Consumer Credit Act 1974) provide that a person wishing to appeal against an order made by the OFT[2] must give notice to the Secretary of State within 28 days;[3] if this notice does not itself specify the grounds of appeal (eg disputed reasons or findings of fact relied on by the OFT, or points of law), then the appellant must serve another notice specifying these grounds before the 28-day period expires. Once the grounds of appeal are made known to the OFT, the latter has 28 days to furnish the Secretary of State with a copy of the original order, together with any representations; the appellant in turn then has 28 days to set out any further representations to be made in the light of the OFT's reply.

Assuming that the Secretary of State is not minded at this stage of the proceedings simply to allow the appeal (or, even if so minded, that the OFT objects to this), the Secretary of State must make arrangements for the appeal to be heard, unless the appellant within

---

1   See s 5(4) and (5).

2   This applies not only to orders under s 3 and s 4, but also to decisions of the OFT under s 6 (revocation and variation of orders) and s 17 (exemption from the insurance requirements of s 16).

3   Compliance with this time-limit is crucial; it is the only one throughout the appeal procedure which the Secretary of Sate is not empowered to extend: reg 26.

14 days requests that the matter be disposed of without a hearing.[1] 21 days' notice of the date, time and place of the hearing must be given to both parties.[2]

Appeals will usually be heard by three 'appointed persons', a legally qualified chairman together with two other members drawn from a panel made up of people with relevant professional experience. While procedure at the hearing is in the hands of the chairman, the regulations clearly envisage that the parties may be represented (by lawyers or others); may call witnesses (who may be questioned by the other party or the panel); and may make opening and closing statements. Further, it is specifically provided that evidence may be admitted whether or not it would be admissible in a court of law.

Once the appointed persons have completed the hearing (or, as the case may be, have considered the matter without a hearing), they are required to furnish the Secretary of State with a reasoned report in writing. On the basis of that report, the Secretary of State notifies the parties of the decision and the reasons for it, together with a reminder to the appellant that a further appeal, on a point of law only, lies to the High Court.

Quite apart from the appeals procedures, a person in respect of whom an order has been made under section 3 or section 4 is permitted by section 6 to make an application to the OFT, asking for the order to be revoked or varied.[3] This provision, which is presumably intended for the benefit of the sinner who repents, may lead to a re-opening of the whole case for, unless the OFT simply accedes to the request, the applicant must be notified of the intention to refuse it (or only to vary the order in a way which does not satisfy the applicant's wishes). Thereafter the position is once again governed by Schedule 2, Part 1 (as to representations, hearings, etc) and by the appeals provisions of section 7.

---

1   The appellant's wishes are not conclusive; the Secretary of State will decide after hearing from the OFT.
2   An appellant who fails to appear runs the risk of the hearing taking place in his or her absence. However, regulations 15 and 17 provide that, in practice, the appellant will be given the opportunity to explain at least a single failure to the Secretary of State.
3   The form of an application, and the conditions on which it may be made, are governed by s 6 and Schedule 2, Part 2. Repeated applications, especially frivolous ones, will no doubt be discouraged by the non-returnable fee involved, which is set at a swingeing £2,500: Estate Agents (Fees) Regulations 1982, SI 1982 No 637.

The OFT, on being asked to revoke an order made under section 3, may refuse if it is considered that the applicant remains unfit within the terms of the order. Similarly, an order made under section 4 will be confirmed if the OFT considers that the applicant is likely to commit further breaches of whatever obligation formed the basis of the order. Where, however, it is felt that the order may, without detriment to the public, be varied in favour of the applicant (eg by restricting it to a particular area or type of work) the OFT may make such a variation.

Register of orders

The essence of the Estate Agents Act is consumer protection and, consistently with this philosophy, section 8 imposes upon the OFT a duty to 'establish and maintain a register on which there shall be entered particulars of every order made by him under section 3 or section 4 above and of his decision on any application for revocation or variation of such an order'. The register states the terms of any order and, where an order is not yet in force, the date on which it will come into operation. If an appeal is pending, the order will nevertheless appear (this is a necessary safeguard for the public) but the pending appeal will also be mentioned. The register is open to public inspection, and any person is entitled to take copies or to ask the OFT for a certified copy of any entry.[1] The latter is regarded as conclusive evidence of the contents of the register; further, any particulars which are entered on the register are presumed to be correct unless and until the contrary is proved.[2]

It is the duty of the OFT to see that the register is kept up to date, and that any change of circumstances is noted. In most cases the OFT will automatically have received the information needed for this task, but could be unaware, for example, that an agent has died or that a relevant period under the Rehabilitation of Offenders Act has been extended. In consequence, while the OFT's duty only arises when it is known that there is something which requires alteration, section 8(3) and Schedule 2, Part 2 ensure that an aggrieved person has the opportunity to bring such matters to the OFT's attention.

---

1   The fees for these services are modest: inspection costs £1 per file, and taking copies works out at approximately 75p per sheet: Estate Agents (Fees) Regulations 1982, SI 1982 No 637.

2   S 8(6).

Investigatory powers

In investigating possible criminal offences under the Estate Agents Act, local weights and measures authorities are able to call upon extensive powers of entry and inspection conferred upon them by section 11. These powers have always been highly controversial, although in practice they have caused very few problems for agents, since the criminal offences concerned relate almost entirely to the handling of clients' money and failure to co-operate with the enforcement authorities, neither of which is an everyday event.

Things may be very different in future, since the present version of section 11[1] provides that local weights and measures authorities may utilise their powers in investigating, not just criminal offences, but also potential breaches of all the other duties created by the 1979 Act and its supporting regulations and orders (including engaging in 'undesirable practices').

The specific powers (all of which are exercisable only by a duly authorised officer of an enforcement authority, only at a reasonable hour and only on production of credentials if demanded) are:

- to enter premises (other than purely residential ones) on reasonable suspicion that an offence has been committed or other breach of duty[2]

- to inspect books or documents, or a legible reproduction of information stored in other forms (eg in a computer) and to take copies, again on reasonable suspicion of an offence or other breach of duty[3]

- on reasonable belief[4] that they are required as evidence in a criminal prosecution, or in proceedings brought by the OFT under sections 3, 4, 6 or 7, to seize and detain books or documents. Where this particular power is exercised, the person from whom

---

1 As amended by the Consumers, Estate Agents and Redress Act 2007, s 57.
2 This provision does not justify a forcible entry; if admission is refused (or in any case where the premises are purely residential), the officer must obtain a warrant from a justice of the peace in accordance with section 11(4).
3 Where a person fails to produce the necessary books, documents or other stored information on demand, the officer can apply for a court order under s 11A.
4 This denotes something more than mere suspicion of an offence.

the books or documents are seized must be informed and, except while they are actually being used as evidence, the person to whom they belong must be given the opportunity to take copies.

The 'teeth' of these provisions are to be found in section 27, which makes it a criminal offence to obstruct an authorised officer who is exercising statutory powers or, in some circumstances, to fail to give the officer adequate assistance in the exercise of those powers. However, no criminal offence is committed where the relevant failure (such as a failure to produce relevant books or documents on demand) can be remedied by a court order under section 11A.

Other Controls on Estate Agency

Miscellaneous statutory provisions

Housing Act 2004

The Home Information Pack scheme

Part 5 of the Housing Act 2004 makes it a requirement, whenever residential property is being marketed, to have in place and to provide to potential buyers a set of documents known as a Home Information Pack (HIP). These documents are designed to speed up the process of residential conveyancing by providing a potential buyer with much of the information that he or she needs before exchange of contracts, and which has traditionally been gathered piecemeal by the buyer's solicitor and other professional advisers. The Act came into force, in relation to properties with at least four bedrooms, on 1 August 2007, and was extended to cover all residential properties from 14 December 2007.

The importance of these provisions for estate agents lies in the fact that, while the statutory duties are imposed upon 'any person who is for the time being responsible for marketing the property',[1] this can only be the seller or a person acting as estate agent for the seller.[2] For this purpose, 'acting as estate agent' means doing anything in pursuance of marketing instructions from the seller; 'marketing instructions' in turn mean instructions to carry out any activities with a view to effecting the introduction to the seller of a person wishing to

---

1   S 151(1).
2   S 151(3).

buy the property or selling the property by auction or tender.[1] The definition is thus similar, though not identical to, that of estate agency work in section 1 of the Estate Agents Act 1979.[2]

Legal obligations in relation to HIPs arise only when a residential property is 'put on the market', that is, when the fact that it is, or may become, available for sale is, with the intention of marketing the property, first made public by or on behalf of the seller.[3] Once this has occurred, the obligations remain in place until the property is taken off the market or sold.

According to section 153, a seller who chooses to market residential property directly[4] is subject to the statutory requirements until such time as he or she appoints an estate agent to handle the marketing and ceases to take any further personal action. More significantly, responsibility is imposed upon the estate agent as soon as he or she takes action to put the property on the market or make public the fact that it is on the market; that responsibility then continues until such time as the agent's contract with the seller is terminated.[5]

Where a residential property is on the market, the person responsible for its marketing (normally an estate agent) is subject to three legal duties. The first and simplest of these is to have under his or her control a home information pack which complies with the relevant regulations. The second is to ensure the authenticity of any document forming part of the home information pack which is shown to a potential buyer.

The third (and most complex) duty is to provide to a potential buyer, on request, a copy of the home information pack, or of a document (or part of a document) included in that pack.[6] This duty does not apply if there is reason to believe that the buyer in question is unlikely to have the means to buy the property, is not genuinely interested in buying that kind of property or is not someone to whom the seller would be prepared to sell the property.[7] The buyer may be

---

1   S 150(1).
2   See p 271.
3   S 149. A fact is 'made public' when it is advertised or otherwise communicated (in whatever form and by whatever means) to the public or a section of the public: s 149(4).
4   This may be a private house-owner or a commercial builder or developer.
5   S 152.
6   An electronic copy is sufficient only with the buyer's consent.
7   However, the seller must not do anything which constitutes unlawful discrimination.

required to pay a charge not exceeding the reasonable cost of making and sending a copy of the relevant documents; the buyer may also be required to accept any terms imposed by the seller as to the use or disclosure of the documents.

The statutory provisions described above apply generally to any residential property[1] which is freehold or subject to a lease for more than 21 years. Importantly, however, section 160 excludes properties which are not sold with vacant possession.[2] A number of other exceptions are contained in regulations; the most important of these are as follows:[3]

- *Non-residential premises.* These are premises where the most recent use was primarily non-residential or which are due to be converted (with all necessary planning consents) to non-residential use. 'Ancillary land' may be treated as non-residential provided it is more than five hectares in extent and was most recently put to agricultural use. However, this exception does not apply to premises which are due to be converted to residential use by the time of completion of the sale.

- *Seasonal and holiday accommodation.* These are premises where a planning condition limits occupation to 11 months in any year or restricts use to holiday accommodation.

- *Mixed sales.* Where a dwelling-house is ancillary to non-residential premises, and it is clear from the way it is marketed that it is not available for separate purchase, the Act does not apply.

- *Dual use.* This refers to property most recently used for both residential and non-residential purposes and marketed in a way that makes clear its suitability for such dual use.

- *Property portfolios.* Such sales are excluded, provided that the seller does not intend to sell any of the properties individually, and the marketing makes this clear.

---

1   That is, a single dwelling-house together with any ancillary land.
2   Where a property for sale is sub-divided into separate dwellings, one or more of which is vacant but none of which is for sale separately (such as a house converted into flats, or a house with a 'granny flat'), the Act *will* apply: s 171.
3   Home Information Pack (No 2) Regulations 2007 (SI 2007 No 1667).

- *Unsafe properties*. The HIP duties do not apply to an unoccupied property whose condition poses a serious risk to health and safety and which is marketed so as to suggest that it is unsuitable for occupation in its present condition.

- *Properties to be demolished*. The Act does not apply to a property whose marketing makes clear that it is suitable for demolition and redevelopment, provided that all relevant planning permissions and consents are in place for this.

Contents of Home Information Packs

Section 163 of the Housing Act empowers the Secretary of State to make regulations prescribing the content of HIPs. The Home Information Pack (No 2) Regulations 2007,[1] as subsequently amended,[2] are the result. These regulations draw a distinction between those documents which are 'required' and those which are merely 'authorised'. The former documents must always be included; the latter are optional. No other document may be included within a HIP, and the regulations specifically forbid the inclusion of any information advertising or marketing goods or services.

The documents which are required are as follows:

- *Index* to the documents contained in the HIP

- *Property Information Questionnaire*. This form, completed by the seller, provides a useful checklist of simple information about the property, of a kind normally found in the answers to standard enquiries raised by a buyer's solicitors.

- *Energy Performance Certificate* or, in the case of a new and as yet uncompleted property, a Predicted Energy Assessment.

---

1   SI 2007 No 1667.
2   See the Home Information Pack (Amendment) Regulations 2007 (SI 2007 No 3301); Home Information Pack (Amendment) Regulations 2008 (SI 2008 No 572); Home Information Pack (Amendment) (No 2) Regulations 2008 (SI 2008 No 1266); Home Information Pack (Amendment) (No 3) Regulations 2008 (SI 2008 No 3107); Home Information Pack (Amendment) Regulations 2009 (SI 2009 No 34).

- *Sustainability information for new homes.* This will be either a certificate showing the sustainability of the property assessed against the standards of the Code for Sustainable Homes, or a nil-rated certificate where the property has not been assessed on this basis.

- *Sale statement.* This provides basic information about the property, such as its address, whether it is freehold, leasehold or commonhold, whether title is registered or unregistered, whether it is to be sold with vacant possession and the capacity in which the seller is acting.

- *Evidence of title* (certain documents obtainable from the Land Registry).

- *Standard searches.* These are searches of the local land charges register and the local authority's records on matters of interest to buyers, such as planning decisions and road building proposals. Also required are details of the provision of drainage and water services to the property.

- In the case of leasehold property, the HIP must also include a copy of the lease.

The first four of these items (from the Index to the evidence on sustainability) must appear in this order and must be the first items in the HIP. The remaining items can appear in any order.

The following documents are authorised, though not required:

- *Home Condition Report.* This report by a qualified and licensed Home Inspector (based on an inspection similar in detail to that of the RICS Homebuyer Survey and Valuation, though not itself including a valuation) was originally intended to be a required document and was indeed the centrepiece of the legislative proposals. However, problems of implementation have led the government to downgrade its status to that of an optional inclusion in the HIP.

- *Non-standard searches,* such as those covering mining, rights of way, ground stability and actual or potential environmental hazards such as flooding and contaminated land.

- *Guarantees or warranties* relating to any work carried out on the property.

- In the case of leasehold property, a range of other information including details of service charges for the last 3 years and a summary of any works either undertaken or proposed which will affect the property.

It is provided by the regulations that all required land registry documentation and searches should be dated no earlier than three months before the first point of marketing, and that all EPCs should be dated no earlier than 12 months before that time. All other documentation should be versions that can reasonably be assumed to be the most recent at the first point of marketing.

Sanctions

Like the Estate Agents Act 1979, Part 5 of the Housing Act 2004 is enforced by the Office of Fair Trading and local weights and measures (trading standards) authorities. The latter have powers under section 167 to call upon an estate agent to produce for inspection a relevant HIP, or document forming part of it, and to take copies. Section 168 makes it a criminal offence to obstruct an authorised officer in this context.

Where an estate agent is in breach of any of the duties imposed by Part 5, this is treated as equivalent to an 'undesirable practice' under section 3(1)(d) of the Estate Agents Act 1979; it is thus a trigger for the powers of the OFT to issue a banning or warning order against the agent. In addition, section 168 and Schedule 8 empower an authorised officer of an enforcement authority to issue a 'penalty charge notice' (effectively a fine to be paid within 28 days), the level of which may not exceed £500.[1] However, where the agent's 'offence' is based on the fact that a document within the HIP does not comply with the statutory requirement, it is a defence for the agent to show that he or she reasonably believed that it did comply.

Where an estate agent fails to provide a potential buyer with a copy of a prescribed document, at a time when the buyer and seller are in negotiations for a sale, there is an additional sanction. The buyer

---

1   Home Information Pack (No 2) Regulations 2007, reg 35.

may commission his or her own copy of the relevant document and recover the cost of so doing from the offending agent.

Accommodation Agencies Act 1953

This Act, which was originally passed as a short-term measure and was due to expire at the end of 1957, was thereafter renewed on an annual basis and was eventually made permanent.[1] None the less, although contraventions of the Act appear to be widespread, these have given rise to legal proceedings on remarkably few occasions. Indeed, it seems that, as Edmund Davies LJ has said: 'The provisions of the 1953 Act are not nearly as well known as they ought to be.'[2]

As a general rule, of course, an estate agent is at liberty to act either for a vendor/landlord or for a purchaser/tenant and, in either case, to charge the appropriate commission. The only restriction placed upon the agent by the general law is not to act for both sides. The effect of the Accommodation Agencies Act is to place a specific limitation upon an agent's right to act on behalf of a prospective tenant.

Section 1 of the 1953 provides as follows:

> (1) Subject to the provisions of this section, any person who, during the continuance in force of this Act,
>
>> (a) demands or accepts payment of any sum of money in consideration of registering, or undertaking to register, the requirements of any person seeking the tenancy of a house;
>>
>> (b) demands or accepts payment of any sum of money in consideration of supplying, or undertaking to supply, to any person addresses or other particulars of houses to let . . . shall be guilty of an offence.[3]

In *McInnes* v *Clarke*[4] the appellant's business was that of finding furnished accommodation for people, in return for a fee equal to one week's rent of the premises taken. His practice was to ask clients how much they were willing to pay as a weekly rental and to take a deposit of approximately half this amount, with an agreement to pay the

---

1   Expiring Laws Continuance Act 1969, section 1.
2   *Crouch & Lees* v *Haridas* [1971] 3 All ER 172, 176.
3   The penalties on summary conviction are a fine of up to one hundred pounds or imprisonment of up to three months, or both: s 1(5).
4   [1955] 1 All ER 346.

balance when accommodation was found. The client then received a list of addresses which was revised daily. A client who accepted one of the properties on the list would be asked to pay the balance of one week's rent. If a client did not find accommodation, the deposit would be refunded on demand. The appellant, who had been convicted on two counts of contravening section 1(1)(b) of the Accommodation Agencies Act, appealed to the Divisional Court of Queen's Bench, arguing that the Act was designed to prevent persons being charged for a list of addresses which might be totally worthless; it did not apply where, as here, the client would ultimately have to pay only in the event of obtaining the desired accommodation.

The Divisional Court dismissed the appeal and held that there had been a clear breach of the Act. Lord Goddard CJ said:

> The statute might have made some provision for the case where a man is honestly conducting a business, but it has not; what it says is, that he must not demand or accept payment of any sum of money in consideration of supplying, or undertaking to supply, to any person addresses or other particulars of houses to let. That is what the appellant did. That was why he received the deposit, and the deposit was to be treated as part payment if the person took one of the houses on the list supplied.

The principle laid down in *McInnes* v *Clarke* led to the conviction of another agent in *Lawrence* v *Sinclair-Taylor*[1] where, once again, a client's deposit could be refunded if no satisfactory accommodation was found. The agent in this case argued that the deposit was in respect of 'services rendered', namely, arranging for the client to view the property and negotiating with the landlord; the Divisional Court, however, agreed with the stipendiary magistrate that the 'substantial consideration' for this payment was supplying the client with the address in question.[2]

In *Saunders* v *Soper*[3] the House of Lords had its first opportunity to consider the effects of the Act. The practice of the agent in that case, who was appealing against two convictions, was not to demand a deposit, but to ask clients to sign an agreement under which a fee would become payable if and when they took a tenancy of

---

1 (1973) 228 EG 1922.
2 A similar decision was reached in *Karim Lalani* v *Birmingham CC* (6 July 1992, unreported).
3 [1975] AC 239.

accommodation found for them by the agent. The House of Lords, while approving the decisions in the two previous cases, regarded them as different in an important respect; as Viscount Dilhorne pointed out:[1]

> As in my view it was not proved in this case that any payment was demanded or accepted for the supplying of an address and as it was proved that the payments were accepted or demanded not for that but for finding [the client] suitable accommodation and due only on her entering into a tenancy, in my opinion she was wrongly convicted.

Thus, it appears, an agent may legitimately charge for finding accommodation for a tenant who actually takes it, but may not demand a fee (even a returnable deposit) for merely supplying a prospective tenant with the address in the first place.

Apart from the criminal liability described above, a breach of the 1953 Act will also render the agency contract illegal. In consequence, the estate agent will forfeit any right to commission and, if money has already changed hands, the client is entitled to recover this. It should be noted that these principles remain valid, notwithstanding that the only reported case in which they have been considered[2] was overruled by the House of Lords in *Saunders* v *Soper*, because it could not be distinguished on the facts.

Control of Advertisements Regulations 2007

The display of advertisements is subject to planning controls under Part VIII of the Town and Country Planning Act 1990. Since an estate agent's 'For Sale' board falls within the statutory definition of 'advertisement', every erection of a board should in principle require a grant of planning permission. Such a situation would of course be unworkable, and the solution lies in the 'deemed consent' provisions of section 222 of the Act. According to that section, planning permission is deemed to have been granted for the display of any advertisement, provided that it complies with regulations made under section 220.[3]

---

1   At p 246.
2   *Crouch & Lees* v *Haridas* [1971] 3 All ER 172, where the Court of Appeal rejected an estate agent's claim for commission for finding an unfurnished flat, of which the client took a five-year lease.
3   If it does not, conviction for an offence under s 224(3) carries, on summary conviction, a fine not exceeding level 3 on the standard scale plus £40 for each day the offence continues after conviction.

The relevant regulations[1] provide that planning permission is deemed to have been granted for the display of:

> [a]n advertisement relating to the sale or letting, for residential, agricultural, industrial or commercial use or for development for such use, of the land or premises on which it is displayed.

There are, however, strict limitations on such matters as the size of a board or its lettering, and a board's positioning (including its projection from a building). It is further specifically provided that illumination is not permitted.

The 'deemed consent' under the 1992 Regulations is designed to permit the advertisement of the property, not the estate agent. Consequently, while 'For Sale' boards are allowed, 'Sold' boards are not. However, an estate agent is specifically permitted to add to an existing board 'a statement that a sale or letting has been agreed, or that the land or premises have been sold or let, subject to contract', provided in any case that 'any such advertisement shall be removed within 14 days after the sale is completed or a tenancy is granted'.

It should be noted that the Regulations limit permission to the display of one advertisement (consisting of a single board or two joined boards). Under earlier regulations, it was argued that an agent who quite lawfully put up a board at a client's property could become guilty of a criminal offence without knowing it if a second agent also erected a board. That argument, which convinced the Queen's Bench Divisional Court, was ultimately rejected by the House of Lords in *Porter* v *Honey*,[2] on the basis that Parliament could not have intended that criminal liability might arise in this way. In any event, the 1992 Regulations have put the matter beyond doubt, by providing specifically that 'where more than one such advertisement is displayed, the first to be displayed shall be taken to be the one permitted'.

Solicitors Act 1974

Section 22 of the 1974 Act is the most recent of a line of statutory provisions in substantially the same form, originating in the Stamp Act

---

1   Town and Country Planning (Control of Advertisements) Regulations 2007 (SI 2007 No 173), reg 6, Sched 3(1) para 1.
2   [1988] 3 All ER 1045.

1804, the general object of which is to give the legal profession a mono-poly in the carrying out of conveyancing work in return for payment.

Section 22(1) provides:

> [A]ny unqualified person who directly or indirectly —
>
> (a) draws or prepares any instrument of transfer or charge for the purposes of the Land Registration Act 1925, or makes any application or lodges any document for registration under the Act at the registry, or
>
> (b) draws or prepares any other instrument relating to real or personal estate, or any legal proceeding,
>
> shall, unless he proves that the act was not done for or in expectation of any fee, gain or reward, be guilty of an offence and liable on summary conviction to a fine not exceeding level 3 on the standard scale.

'Unqualified' for this purpose means anyone who is not a solicitor, although there are a few exemptions, of which the most important are barristers and notary publics,[1] licensed conveyancers[2] and, in relation to farm business tenancies, members of the Royal Institution of Chartered Surveyors or the Central Association of Agricultural Valuers.[3]

As originally enacted, the prohibition in this section was limited by its definition of 'instrument', which covered only agreements made by deed. In relation to the transfer of freehold property, this meant that an unqualified person such as an estate agent would be guilty of a criminal offence in preparing a deed of conveyance, but could legitimately draw up a contract of sale. However, this opening was firmly closed by an amendment introduced in 1985,[4] as a result of which section 22(3) now provides:

> For the purposes of subsection (1)(b), 'instrument' includes a contract for the sale or other disposition of land (except a contract to grant such a lease as is referred to in section 54(2) of the Law of Property Act 1925 (short leases)), but does not include —
>
> ...
>
> (b) an agreement not intended to be executed as a deed other than a contract that is included by virtue of the preceding provisions of this subsection.

---

1   S 22(2).
2   Administration of Justice Act 1985, s 11(4).
3   Agricultural Tenancies Act 1995, s 35(2) and (3).
4   Administration of Justice Act 1985, s 6(4).

In relation to the creation of leases, estate agents have attempted to evade prosecution under earlier statutes by preparing the relevant document, but then not executing it as a deed. Before section 1 of the Law of Property (Miscellaneous Provisions) Act 1989 came into force, this was achieved by not affixing a seal to the document; in the light of that Act, however, the equivalent is simply to ensure that the document does not state that it is to take effect as a deed. In any event, such attempts have mostly proved ineffective, as illustrated by the case of *Harte* v *Williams*.[1] An estate agent there, acting on behalf of a landlord, found a person willing to take a 21-year lease of property. The agent then offered to prepare the lease on behalf of the prospective tenant, in order to save him the expense of a solicitor's fee. When this offer was accepted, the estate agent filled in a standard-form document commonly used by solicitors and had a number of copies made; no seal, however, was affixed, although the law at that time required one for a lease of this length. When the agent charged a small fee for this service of 'preparing copy lease', he was convicted of an offence under the Stamp Act 1891, section 44. On appeal to a Divisional Court of King's Bench, it was strongly argued that the document in question was 'an agreement under hand only' (ie not made by deed) but a majority of the court rejected this, preferring the view that 'an agreement under hand only means an instrument valid and intended to operate in law as such'. What the agent had prepared did not satisfy this definition; it was merely a lease which was void because it did not take the form of a deed.

Of course, there are some leases which may be validly created without using a deed and, where this is so, an estate agent may lawfully charge for preparing the relevant document. It is for this reason that section 22(3) refers specifically to section 54(2) of the Law of Property Act 1925, under which a deed is not required for the creation of a lease which 'takes effect in possession for a term not exceeding three years (whether or not the lessee is given power to extend the term) at the best rent which can be obtained without taking a fine'.

A periodic tenancy (monthly, yearly, etc) is within this definition, since it may well not last for three years;[2] however, a lease for more than three years must be made by deed, notwithstanding that it may be determined at an earlier date. In *Kushner* v *Law Society*[3] an estate

---

1   [1934] 1 KB 201.
2   *Re Knight, ex p Voisey* (1882) 21 ChD 442.
3   [1952] 1 KB 264.

agent, in return for a fee from the tenant, drew up a document which purported to grant him the tenancy of a flat for 14 years, with a provision that the tenant might terminate the agreement by notice at the end of any year. A Divisional Court of King's Bench unanimously held that the estate agent was rightly convicted of an offence under the Solicitors Act 1932, section 47; since the lease was one which could only validly be made by deed, the case was identical to that of *Harte* v *Williams*. As Lord Goddard CJ said:[1]

> The offence consists not in procuring the execution of a document, but in drawing or preparing it ... A person who draws or prepares a document which if it is to be a valid document, must have a seal on it at the time of execution, commits an offence; because he has drawn the deed he commits the offence none the less although the document when executed does not have a seal put upon it.

The Ombudsman for Estate Agents scheme

During the 1980s, there was a groundswell of consumer criticism of the activities of, and the services provided by, some 'corporate' estate agents, that is, those agency chains owned by large institutions (mainly banks and building societies). In September 1990, as a response to this criticism, the institutions concerned set up the Ombudsman for Corporate Estate Agents (OCEA) Scheme. Under this Scheme, complaints against estate agencies owned by these institutions were to be handled, without charge to the complainant, by an independent person, whose decision (which might include the award of financial compensation to the complaint) they undertook to honour. By the end of 1997, the Scheme was in operation in the estate agency chains owned by ten corporate members, a total of some 2,000 offices.

On 1 January 1998, following discussions between the Member Agencies and the three main professional bodies representing estate agents,[2] the OCEA Scheme was replaced by the Ombudsman for Estate Agents (OEA) Scheme, membership of which was open to any

---

1   At p 271.
2   These were the Royal Institution of Chartered Surveyors, the Incorporated Society of Valuers and Auctioneers (which ceased separate existence on 1 January 2000, on its merger with the RICS) and the National Association of Estate Agents.

estate agency firm in which a principal, partner or director was a member of one of the professional bodies. From 1 January 2000, the OEA Scheme was further widened, becoming open to any estate agent who was prepared to sign up to all the commitments required of members, and who was proposed and seconded by existing members.

By 2004, some 40 per cent of estate agency offices in the United Kingdom had joined the OEA Scheme. However, the government's desire to introduce compulsory membership of an 'approved' redress scheme (initially floated for those agents handling Home Information Packs by the Housing Act 2004) was applied to all residential estate agents by the Consumers, Estate Agents and Redress Act 2007, which came into force on 1 October 2008. At the time of writing, two schemes had been approved for this purpose; the OEA Scheme and the one administered by Surveyors Ombudsman Services Ltd for agents who are RICS members, and it is estimated that some 90 per cent of residential agents are now members of the OEA Scheme.

Jurisdiction of the Ombudsman for Estate Agents

Under the OEA Scheme, the Ombudsman is empowered to investigate complaints made, against estate agents who are members of the Scheme, by private individuals who are actual or potential sellers or buyers of residential property in the United Kingdom.[1] Such complaints may be based on any of the following grounds:

- Infringement of the complainant's legal rights;
- Failure to comply with the OEA Code of Practice;[2]
- Unfair treatment of the complainant;
- Maladministration (including inefficiency or undue delay).

In all these cases, the complainant must show that he or she has lost money or suffered inconvenience as a result of the estate agent's conduct. If this requirement is satisfied, the Ombudsman is empowered to make an award of financial compensation, up to a maximum of £25,000.[3] The member agency is bound to honour such an

---

1   The Scheme does not apply to complaints brought by one estate agent against another.
2   See below.
3   The highest award to date is £13,100, and most awards are for less than £500.

award, provided that the complainant agrees to accept it in full and final settlement of the claim.

A number of matters are specifically excluded from the Ombudsman's jurisdiction. Of these, the most important are complaints arising out of surveys or formal valuations of property, or out of lettings or property management.[1] In addition, the Scheme is clearly intended to operate in lieu of other mechanisms for seeking redress; thus the Ombudsman cannot deal with any complaint which is, or has been, handled by a court or other similar tribunal.

The Ombudsman will almost always refuse to deal with a complaint, until such time as the complainant has exhausted the in-house complaints procedure of the member agency concerned. This requirement brings into play two important time-limits. The Ombudsman will not act where the complainant delayed more than 12 months before submitting a written complaint to the member agency; nor where the complainant has waited more than six months from the member agency's final offer of settlement before submitting a complaint to the Ombudsman.

Procedures and outcomes

Where a private individual has been through a member agent's internal complaints procedure but is dissatisfied with the outcome, he or she can make a written complaint to the Ombudsman (the OEA Scheme provides a form). The complainant is then asked to sign a letter waiving the right to confidentiality, which enables the Ombudsman to obtain the relevant file from the estate agency concerned. When sending the file to the Ombudsman, the agency encloses a letter explaining how it has dealt with the dispute and presenting its side of the case. All the documents are then examined by the Ombudsman's staff, whose initial findings are reviewed by the Ombudsman personally.

Assuming that the case is thought suitable for resolution under the Scheme,[2] the Ombudsman may make one of three decisions. The

---

1   Also excluded are services provided on special terms by member agents to employees or spouses of employees.

2   Very occasionally (for instance where there is an irreconcilable conflict of evidence) the Ombudsman may take the view that the case is better suited for resolution by a court, where the witnesses can give evidence, and be cross-examined on it, under oath.

first of these is a finding in favour of the complainant, in which case a draft of the proposed award is sent, first to the estate agent (for comments and a possible appeal) and then to the complainant (for comments, a possible appeal and, ultimately, acceptance).[1] Second, the Ombudsman may reject the complaint, in which case the complainant is informed that, unless further evidence can be provided, the matter is closed. Finally, where the estate agent's internal complaints procedure has resulted in a 'without prejudice' offer of settlement, but the complainant has rejected this, the Ombudsman may reach a decision as to whether the amount offered is appropriate.

The OEA Code of Practice

One of the central features of the OEA Scheme is the Code of Practice for Residential Estate Agents, compliance with which is mandatory for member agents and for whose breach the Ombudsman is empowered to award damages to a complainant. The Code operates to reinforce the obligations imposed upon estate agents by the general law, by specifically requiring compliance with relevant legislation. However, it goes further by creating some general obligations, such as 'to provide a service consistent with fairness, integrity and best practice' and some more specific obligations, such as protecting a client's confidential information and warning a buyer if he becomes involved in a contract race.

The original OEA Code of Practice was drafted in consultation with the main professional bodies representing estate agents. The present version, which is both more detailed and wider in scope, was drafted by the OEA alone and has received the approval of the OFT under its Consumer Codes and Approval Scheme.

This is not the place for a detailed examination of the Code of Practice for Residential Estate Agents. However, some idea of its coverage may be gleaned from the headings and sub-headings under which its provisions are grouped:

1. General — compliance with all relevant legislation; responsibility for the conduct and awareness of all agency staff; provision of a

---

1   The 'appeals' mentioned are in effect requests to the Ombudsman to look again at the matter, rather than appeals to a separate body to review the decision. They will normally succeed only where the Ombudsman has made a significant error of fact, or where new evidence is produced.

service consistent with fairness, integrity and best practice; non-discrimination; fairness to vulnerable consumers; confidentiality; record-keeping.

2. Market Appraisal — all advice on price to be honest, reflective of market conditions and, where possible, supported by comparable evidence.

3. Terms of Business, Instructions, Commission and Termination — no harassment in order to gain instructions; no instruction of a sub-agent without the client's consent; terms of business to be in plain and intelligible language; fees to be clearly stated; provision for termination of agency; warning to client of possibility of dual fees; terms to be signed and dated.

4. Marketing and Advertising — compliance with legislation on sale boards; no board to be erected without seller's specific permission; all advertisements to be legal, decent, honest and truthful.

5. Viewing and Access to Property — all viewings in accordance with seller's instructions; records to be kept; duty of care where in possession of keys to property.

6. Submission of Offers — information to seller on offers, and records kept; information to buyers where property withdrawn from market; transparency towards all parties in negotiations.

7. Financial Evaluation — duty to take reasonable steps to check source and availability of buyer's funds.

8. Deposits — no pre-contract deposits except on new homes; no deposits unless bonded.

9. Duty of Care and Conflict of Interest — duty of care to seller; fair and courteous treatment of all parties; avoidance of conflicts of interest; non-intimidation in pursuit of fees.

10. Between Acceptance and Exchange of Contracts — assistance where possible to seller with conveyancing process; warning to buyer of contract race.

11. Exchange and Completion — proper handling of keys to property.

12. In-house Complaints Handling — maintenance and minimum requirements of proper in-house procedure

13. Referrals to the Ombudsman — co-operation with the Ombudsman and compliance with any award.

14. Compliance Monitoring — compliance with OEA procedures to monitor compliance with the Code of Practice.

Index